SpringerBriefs in Mathematics

SpringerBriefs present concise summaries of cutting-edge research and practical applications across a wide spectrum of fields. Featuring compact volumes of 50 to 125 pages, the series covers a range of content from professional to academic. Briefs are characterized by fast, global electronic dissemination, standard publishing contracts, standardized manuscript preparation and formatting guidelines, and expedited production schedules.

Typical topics might include:

- A timely report of state-of-the art techniques
- A bridge between new research results, as published in journal articles, and a contextual literature review
- A snapshot of a hot or emerging topic
- An in-depth case study
- A presentation of core concepts that students must understand in order to make independent contributions

SpringerBriefs in Mathematics showcases expositions in all areas of mathematics and applied mathematics. Manuscripts presenting new results or a single new result in a classical field, new field, or an emerging topic, applications, or bridges between new results and already published works, are encouraged. The series is intended for mathematicians and applied mathematicians. All works are peer-reviewed to meet the highest standards of scientific literature.

Titles from this series are indexed by Scopus, Web of Science, Mathematical Reviews, and zbMATH.

Yuefang Sun

Steiner Type Packing Problems in Digraphs

Yuefang Sun
School of Mathematics and Statistics
Ningbo University
Ningbo, Zhejiang, China

ISSN 2191-8198 ISSN 2191-8201 (electronic)
SpringerBriefs in Mathematics
ISBN 978-981-95-8742-1 ISBN 978-981-95-8743-8 (eBook)
https://doi.org/10.1007/978-981-95-8743-8

This Springer imprint is published by the registered company Springer Nature Singapore Pte Ltd.
The registered company address is: 152 Beach Road, #21-01/04 Gateway East, Singapore 189721, Singapore

For my family

Preface

Graph packing problem is one of the central problems in graph theory and combinatorial optimization. The famous Steiner tree packing problem in undirected graphs is not only an important graph-theoretical problem but also has a strong background in applications, especially in VLSI circuit design and Internet Domain. It attracts much attention of researchers in the areas of graph theory, combinatorial optimization, and theoretical computer science, and has become a well-established area. It is natural to extend this problem to digraphs, and such problems in digraphs are called *Steiner type packing problems in digraphs*. There are several Steiner type packing problems in digraphs: directed Steiner tree packing problem, directed Steiner path packing problem, strong subgraph packing problem, strong arc decomposition problem, and directed Steiner cycle packing problem.

An *out-tree* is an oriented tree in which every vertex except one, called the *root*, has in-degree one. An *out-branching* of D is a spanning out-tree in D. For a digraph $D = (V(D), A(D))$, and a set $S \subseteq V(D)$ with $r \in S$ and $|S| \geq 2$, a *directed* (S, r)*-Steiner tree* or, simply, an (S, r)*-tree* is an out-tree T rooted at r with $S \subseteq V(T)$. Two (S, r)-trees are said to be *arc-disjoint* if they have no common arc. Two arc-disjoint (S, r)-trees are said to be *internally-disjoint* if the set of common vertices of them is exactly S. Let $\kappa_{S,r}(D)$ (resp. $\lambda_{S,r}(D)$) be the maximum number of internally-disjoint (resp. arc-disjoint) (S, r)-trees in D. The problem of ARC-DISJOINT DIRECTED STEINER TREE PACKING (ADSTP) is defined as follows: the input consists of a digraph D and a subset of vertices $S \subseteq V(D)$ with a root r, the goal is to find a largest collection of arc-disjoint (S, r)-trees. When we replace "arc-disjoint" by "internally-disjoint" in ADSTP, we get the definition of the problem of INTERNALLY-DISJOINT DIRECTED STEINER TREE PACKING (IDSTP). It is worth mentioning that the problem of directed Steiner tree packing is highly related to two important theorems in graph theory: when $|S| = 2$, that is, $S = \{r, x\} \subseteq V(D)$, then $\lambda_{S,r}(D) = \lambda_D(r, x)$ (resp. $\kappa_{S,r}(D) = \kappa_D(r, x)$), the local arc-strong (resp. vertex-strong) connectivity of r, x in D, and therefore the problem is related to the famous Menger's Theorem; when $|S| = n$, the problem is equivalent to finding a largest collection of arc-disjoint out-branchings rooted at r, and therefore is related to the famous Edmonds' Branching Theorem. In the definition of ADSTP, if each (S, r)-tree is a directed path, then the problem is called ARC-DISJOINT DIRECTED STEINER PATH PACKING (ADSPP). Similarly, we can define the problem of INTERNALLY-DISJOINT DIRECTED STEINER PATH PACKING (IDSPP).

Let $D = (V(D), A(D))$ be a digraph of order n, S a k-subset of $V(D)$ and $2 \leq k \leq n$. A strong subgraph H of D is called an *S-strong subgraph* if $S \subseteq V(H)$. Two S-strong subgraphs are said to be *arc-disjoint* if they have no common arc. Furthermore, two arc-disjoint S-strong subgraphs are said *internally-disjoint* if the set of common vertices of them is exactly S. We use $\kappa_S(D)$ (resp. $\lambda_S(D)$) to denote the maximum number of internally-disjoint (resp. arc-disjoint) S-strong subgraphs in D. The problem of ARC-DISJOINT STRONG SUBGRAPH PACKING (ASSP) is defined as follows: the input consists of a digraph D and a subset of vertices $S \subseteq V(D)$, the goal is to find a largest collection of arc-disjoint S-strong subgraphs. When we replace "arc-disjoint" by "internally-disjoint" in ASSP, we get the definition of the problem of INTERNALLY-DISJOINT STRONG SUBGRAPH PACKING (ISSP). In the definition of ASSP, if each S-strong subgraph is a directed cycle, then the problem is called ARC-DISJOINT DIRECTED STEINER CYCLE PACKING (ADSCP). Similarly, we can define the problem of INTERNALLY-DISJOINT DIRECTED STEINER CYCLE PACKING (IDSCP). Observe that the problem of directed Steiner cycle packing is a generalization of the famous directed Hamiltonian cycle decomposition problem. A digraph $D = (V(D), A(D))$ has a *strong arc decomposition* if A has two disjoint sets A_1 and A_2 such that both $(V(D), A_1)$ and $(V(D), A_2)$ are strong. Clearly, a digraph D has a strong arc decomposition if and only if $\lambda_{V(D)}(D) \geq 2$ (or, $\kappa_{V(D)}(D) \geq 2$). Therefore, the problem of ASSP (or ISSP) could be seen as an extension of the strong arc decomposition problem.

In this book, we try to bring together most of known results on Steiner type packing problems in digraphs. We begin with an introductory chapter which includes some terminology and notion in digraph theory and computational complexity theory, and the backgrounds of undirected Steiner tree packing problem and directed Steiner type packing problems. In Chap. 2, we mainly introduce results for directed Steiner tree packing problem, including the complexity for the decision versions of IDSTP and ADSTP on general digraphs, symmetric digraphs, and Eulerian digraphs. Inapproximability results on ADSTP and IDSTP and their generalizations are also mentioned. In the last part of this chapter, we introduce three related topics: directed tree connectivity, independent directed Steiner trees and independent branchings, and arc-disjoint in- and out-branchings rooted at the same vertex. In Chap. 3, we introduce the algorithmic and complexity results for the directed Steiner path packing problem. In the last part of this chapter, the topic of directed path connectivity is also mentioned. In Chap. 4, we mainly introduce results for strong subgraph packing problem, including the complexity for the decision versions of ISSP or/and ASSP on general digraphs, semicomplete digraphs, symmetric digraphs, and Eulerian digraphs. Inapproximability results on ISSP and ASSP are also mentioned. The last part of this chapter is about two related topics. The first one includes Kriesell conjecture and its extension in strong subgraph packing problem, and the second one concerns another connectivity on digraphs, strong subgraph connectivity. Chapter 5 concerns the strong arc decomposition of semicomplete digraphs, locally semicomplete digraphs, digraph compositions, and digraph products. In this chapter, we also introduce Kelly conjecture and Bang-Jensen–Yeo conjecture which are highly related to strong arc decomposition. In the final chapter, Chap. 6, we introduce the algorithmic and complexity results for the directed Steiner cycle packing problem, and a related topic: directed cycle k-connectivity. In each chapter we list conjectures or open problems at appropriate places. We hope that this can motivate more young researchers and graduate students to do further study in this subject. We do not give proofs for all results. Instead, we only select some of them for which we give their proofs because we feel that these proofs employed some typical techniques which are popular in the study of directed Steiner type packing problems. New results are still appearing, there must be some or even many of them for which we have not realized their existence, and therefore have not included them in this book.

The readers of the book are expected to have some backgrounds in graph theory and some related knowledge in algorithms and complexity analysis. All relevant terminology and notation from graph theory and computational complexity theory are properly defined in Chap. 1, but also elsewhere where needed.

The anticipated readers of the book are mathematicians and students of mathematics, whose fields of interest are graph theory, combinatorial optimization, theoretical computer science as well as communication network design. Consequently, the present book will be found suitable for courses in these fields. The exposition of the details of the proofs of some main results will enable students to understand and eventually master a good part of graph theory, combinatorial optimization and theoretical computer science. People working on communication networks may also be interested in some aspects of the book. They will find it useful for designing networks that can efficiently, reliably, and safely transfer information.

The material presented in this book was used in digraph theory seminars, held three times at School of Mathematics and Statistics, Ningbo University, in 2022 and 2024. I thank all the members of my group, especially Yiling Dong, Ruixiao Jing, Ansong Ma, Chuchu Wang, Jun Wang, Xiusen Wang, Xiaosha Wei, Mingxia Wu, Junran Yu, Shanshan Yu, and Yangfan Yu, for help in the preparation of this book. I also thank the National Natural Science Foundation of China under Grant No. 12371352 for financial support to our research project "Computational Complexity and Sufficient Conditions of Directed Steiner Type Packing Problems," the Zhejiang Provincial Natural Science Foundation of China under Grant No. LY23A010011 for financial support to our research project "Some Topics on Directed Steiner Tree Packing," and the Yongjiang Talent Introduction Programme of Ningbo under Grant

No. 2021B-011-G for financial support to our research project "Digraph Packing." Last but not least, I am very grateful to Associate Editor of Mathematics & Statistics of Springer Nature, Daniel Wang. Without his encouragement, this book may not exist.

Ningbo, China Yuefang Sun

Contents

Chapter 1
Introduction

Abstract In this chapter, we first introduce terminology and notation in digraph theory and computational complexity theory. Then we introduce the backgrounds of Steiner tree packing problem in undirected graphs, and several Steiner type packing problems in digraphs.

1.1 Terminology and Notation

1.1.1 Digraph Theory

We refer the readers to [5] for graph theoretical notation and terminology not given here. Throughout this book, unless otherwise stated, paths and cycles are always assumed to be directed, and all digraphs considered in this paper have no parallel arcs or loops. We use $[n]$ to denote the set of all natural numbers from 1 to n. For a set S, we use $|S|$ to denote the number of elements in S.

Let $D = (V(D), A(D))$ be a digraph. If $xy \in A(D)$, then we say that x *dominates* y and write $x \rightarrow y$ to indicate this. Two vertices x and y are *adjacent* if there is at least one arc between them. If $xy \in A(D)$, then x is called an *in-neighbour* of y, and y is called an *out-neighbour* of x. We use $N_D^+(x)$ (resp. $N_D^-(x)$) to denote the set of all out-neighbours (resp. in-neighbours) of x in D. Let $d^+(x) = |N_D^+(x)|$ (resp. $d^-(x) = |N_D^-(x)|$) denote the *out-degree* (resp. *in-degree*) of x in D. The *degree* $d(x)$ of a vertex x is defined as $d(x) = d^+(x) + d^-(x)$. If we want to specify the degree of a vertex in a subdigraph D' of D, then we may write $d_{D'}^+(x)$. The *minimum semi-degree* of a digraph D, denoted $\delta^0(D)$, is $\delta^0(D) = \min_{x \in V}\{\min\{d^+(x), d^-(x)\}\}$.

If A and B are disjoint subsets of $V(D)$ such that there is no arc from B to A and $a \rightarrow b$ for all $a \in A$ and $b \in B$, then we say that A *(completely) dominates* B and denote this by $A \Rightarrow B$. Let $S \subseteq V(D)$, we use $d^+(S)$ to denote the number of arcs from S to $V(D) \setminus S$ and use $d^-(S)$ to denote the number of arcs from $V(D) \setminus S$ to S.

A digraph D is *connected* if its underlying undirected graph (that is, an undirected graph obtained from D by deleting the direction of each arc and possible

Y. Sun, *Steiner Type Packing Problems in Digraphs*, SpringerBriefs in Mathematics,
https://doi.org/10.1007/978-981-95-8743-8_1

multiple edges) is connected. A digraph D is *strongly connected* (or, *strong*), if for any pair of vertices $x, y \in V(D)$, there is a path from x to y in D, and vice versa. A subdigraph H of D is called a *strong component* if it is a maximal strong connected subdigraph (or, strong subgraph for short) of D. If D is strong and $S \subset V(D)$ such that $D - S$ is not strong, then S is a *separating set*. A separating set S is *minimal* if no proper subset of S is a separating set of D. For any non-strong digraph D, we can label its strong components $D_1, D_2, \ldots, D_s$, $s \geq 2$, in such a way that there is no arc from D_j to D_i when $j > i$. We call this an *acyclic ordering* of the strong components of D.

An *out-tree* (resp. *in-tree*) is an oriented tree in which every vertex except one, called the *root*, has in-degree (resp. out-degree) one. An *out-branching* (resp. *in-branching*) of D is a spanning out-tree (resp. in-tree) in D.

A digraph D is *semicomplete* if for every distinct $x, y \in V(D)$ at least one of the arcs xy, yx is in D. A digraph D is *locally semicomplete* if $N^-(x)$ and $N^+(x)$ induce semicomplete digraphs for every vertex x of D. A semicomplete digraph without 2-cycles is a *tournament* and a locally semicomplete digraph without 2-cycles is a *local tournament*. A digraph is *locally in-semicomplete* (resp. *locally out-semicomplete*) if $N^-(x)$ (resp. $N^+(x)$) induces a semicomplete digraph for every vertex x of D. A digraph D is *quasi-transitive*, if for any triple x, y, z of distinct vertices of D, if xy and yz are arcs of D then either xz or zx or both are arcs of D. A digraph D is *symmetric* if $xy \in A(D)$ then $yx \in A(D)$. In other words, a symmetric digraph D can be obtained from its underlying undirected graph G by replacing each edge of G with the corresponding arcs of both directions, that is, $D = \overleftrightarrow{G}$. For a digraph D, its *reverse* D^{rev} is a digraph with the same vertex set such that $xy \in A(D^{\mathrm{rev}})$ if and only if $yx \in A(D)$. Clearly, a digraph D is symmetric if and only if $D^{\mathrm{rev}} = D$. A digraph is *k-regular* if $d^+(x) = d^-(x) = k$ for every vertex x. A digraph is *Eulerian* if D is connected and $d^+(x) = d^-(x)$ for every vertex x. Observe that a connected symmetric digraph is Eulerian. A digraph is *planar* if it can be drawn on a plane without arc crossings. A digraph is *Hamiltonian decomposable* if it has a family of Hamiltonian cycles such that every arc of the digraph belongs to exactly one of the cycles. We use $\overline{K}_n$ to stand for the digraph of order n with no arcs. Also, $\overrightarrow{C}_n$ and $\overrightarrow{P}_n$ will denote the cycle and path with n vertices, respectively. The *second power* of a cycle $\overrightarrow{C}_n$, denoted by $\overrightarrow{C}_n^2$, is the digraph obtained from $\overrightarrow{C}_n$ by adding the arcs $\{v_i v_{i+2} \mid i \in [n]\}$, where $\overrightarrow{C}_n = v_1 v_2 \ldots v_{n-1} v_n v_1$ and subscripts are modulo n. Furthermore, we can define the *kth power* of a cycle $\overrightarrow{C}_n$ which is denoted by $\overrightarrow{C}_n^k$.

A digraph D has a *strong arc decomposition* if $A(D)$ can be partitioned into two subsets A_1 and A_2 such that both $D_1 = (V(D), A_1)$ and $D_2 = (V(D), A_2)$ are strong.

An *ear decomposition* of a digraph D is a sequence $\mathcal{P} = (P_0, P_1, P_2, \ldots, P_t)$, where P_0 is a cycle or a vertex, and each P_i is a path or a cycle with the following properties:

(a) P_i and P_j are arc-disjoint when $i \neq j$.

(*b*) For each $i = 0, 1, 2, \dots, t$, let D_i denote the digraph with vertices $\bigcup_{j=0}^{i} V(P_j)$ and arcs $\bigcup_{j=0}^{i} A(P_j)$. If P_i is a cycle, then it has precisely one vertex in common with $V(D_{i-1})$. Otherwise the end vertices of P_i are distinct vertices of $V(D_{i-1})$ and no other vertex of P_i belongs to $V(D_{i-1})$.
(*c*) $\bigcup_{j=0}^{t} A(P_j) = A(D)$.

Let $V' \subseteq V(D)$. A subset $S \subseteq V(D) \setminus V'$ is *V'-normal* if every directed walk which leaves and again enters S must contain a vertex of V'. For any two vertices r, r' of an out-tree T, we say $r \leq r'$ if there is an (r, r')-path or $r = r'$. An *arboreal decomposition* of D is a triple (T, X, W), where T is an out-tree (not a subdigraph of D), $X = \{X_e \mid e \in A(T)\}$ and $W = \{W_r \mid r \in V(T)\}$ are sets of vertices of D that satisfy the following two conditions:

(*a*) $W = \{W_r \mid r \in V(T)\}$ is a partition of $V(D)$ such that each W_r is nonempty.
(*b*) $\bigcup\{W_r \mid r \in V(T), r \geq r''\}$ is X_e-normal for each $e = r'r'' \in A(T)$.

The *width* of (T, X, W) is the least integer w such that $|W_r \cup \bigcup_{e \sim r} X_e| \leq w + 1$ for each $r \in V(T)$, where $e \sim r$ means that r is incident to e. The *directed tree-width* of D, denoted by $dtw(D)$, is the least integer w such that D has an arboreal decomposition of width w.

Let T be a digraph with vertices $u_1, \dots, u_t$ $(t \geq 2)$ and let $H_1, \dots, H_t$ be digraphs such that H_i has vertices u_{i,j_i}, where $j_i \in [n_i]$. Let $n_0 = \min\{n_i \mid i \in [t]\}$. Then the *composition* $Q = T[H_1, \dots, H_t]$ is a digraph with vertex set

$$V(Q) = \{u_{i,j_i} \mid i \in [t], j_i \in [n_i]\}$$

and arc set

$$A(Q) = \cup_{i=1}^{t} A(H_i) \cup \{u_{i,j_i} u_{p,q_p} \mid u_i u_p \in A(T), j_i \in [n_i], q_p \in [n_p]\}.$$

The composition $Q = T[H_1, \dots, H_t]$ is *proper* if $Q \neq T$, i.e. at least one H_i is nontrivial, and is *semicomplete* if T is semicomplete. If $Q = T[H_1, \dots, H_t]$ and none of the digraphs $H_1, \dots, H_t$ has an arc, then Q is an *extension* of T. For any set Φ of digraphs, Φ^{ext} denotes the (infinite) set of all extensions of digraphs in Φ, which are called *extended Φ-digraphs*. For example, if Φ is the set of all semicomplete digraphs, then Φ^{ext} denotes the set of all extended semicomplete digraphs.

A digraph on n vertices is *round* if we can label its vertices $u_1, \dots, u_n$ such that for each i, we have $N^+(u_i) = \{u_j \mid i + 1 \leq j \leq i + d^+(u_i)\}$ and $N^-(u_i) = \{u_j \mid i - d^-(u_i) \leq j \leq i - 1\}$ (all subscripts are taken modulo n). Note that every round digraph is locally semicomplete. A locally semicomplete digraph D is *round decomposable* if there exists a round local tournament R on $r \geq 2$ vertices such that $D = R[S_1, \dots, S_r]$, where each S_i is a strong semicomplete digraph. We call $R[S_1, \dots, S_r]$ a *round decomposition* of D.

We now introduce the definitions of several product digraphs. The *Cartesian product* $G \Box H$ of two digraphs G and H is a digraph with vertex set

$$V(G\square H) = V(G) \times V(H) = \{(x, x') \mid x \in V(G), x' \in V(H)\}$$

and arc set

$$A(G\square H) = \{(x, x')(y, y') \mid xy \in A(G), x' = y', \ or \ x = y, x'y' \in A(H)\}.$$

We define the *kth power* with respect to Cartesian product as $D^{\square k} = \underbrace{D\square D\square \cdots \square D}_{k \text{ times}}$.

The *strong product* $G \boxtimes H$ of two digraphs G and H is a digraph with vertex set

$$V(G \boxtimes H) = V(G) \times V(H) = \{(x, x') \mid x \in V(G), x' \in V(H)\}$$

and arc set

$$A(G \boxtimes H) = A(G\square H) \cup \{(x, x')(y, y') \mid xy \in A(G), \ x'y' \in A(H)\}.$$

The *lexicographic product* $G \circ H$ of two digraphs G and H is a digraph with vertex set

$$V(G \circ H) = V(G) \times V(H) = \{(x, x') \mid x \in V(G), x' \in V(H)\}$$

and arc set

$$A(G \circ H) = \{(x, x')(y, y') \mid xy \in A(G), \ or \ x = y \ and \ x'y' \in A(H)\}.$$

1.1.2 Computational Complexity Theory

We still need some basic terminology and notation in computational complexity theory. A *decision problem* is a problem whose answer is either "yes" or "no". Such a problem belongs to the class $\mathcal{P}$ if there is a polynomial-time algorithm that solves any instance of this problem in polynomial time. It belongs to the class $\mathcal{NP}$ if, given any instance of the problem whose answer is "yes", there is a certificate validating this fact, which can be checked in polynomial time. Such a certificate is said to be *succinct*. It is immediate from these definitions that $\mathcal{P} \subseteq \mathcal{NP}$, as a polynomial-time algorithm itself is exactly a succinct certificate. A *polynomial reduction* of a problem P to another problem Q is a pair of polynomial-time algorithms, one of which transforms each instance I of P to an instance J of Q and the other of which transforms a solution for the instance J to a solution for the instance I. If such a reduction exists, we say that P is polynomially reducible to Q, denoted by $P \preceq Q$. A problem $P \in \mathcal{NP}$ is *NP-complete* if $P' \preceq P$ for every problem $P' \in \mathcal{NP}$. For more information on computational complexity theory, the readers can see [28].

In the proofs of complexity results on directed Steiner type packing problems, we will use some complexity results of the following problems:

Directed k-Linkage: For a fixed integer $k \geq 2$, given a digraph D and a (terminal) sequence $((s_1, t_1), \ldots, (s_k, t_k))$ of distinct vertices of D, decide whether D has k vertex-disjoint paths $P_1, \ldots, P_k$, where P_i starts at s_i and ends at t_i for all $i \in [k]$.
Directed Weak k-Linkage: For a fixed integer $k \geq 2$, given a digraph D and a (terminal) sequence $((s_1, t_1), \ldots, (s_k, t_k))$ of distinct vertices of D, decide whether D has k arc-disjoint paths $P_1, \ldots, P_k$, where P_i starts at s_i and ends at t_i for all $i \in [k]$.
CLLM Problem: Given a tripartite graph $G = (V, E)$ with a 3-partition (A, B, C) such that $|A| = |B| = |C| = q$, decide whether there is a partition of V into q disjoint 3-sets $V_1, \ldots, V_q$ such that for every $V_i = \{a_{i_1}, b_{i_2}, c_{i_3}\}$, $a_{i_1} \in A, b_{i_2} \in B, c_{i_3} \in C$ and $G[V_i]$ is connected.
3-SAT: Given a Boolean formula ϕ in conjunctive normal form with three literals per clause, decide whether ϕ is satisfiable.
2-Coloring Hypergraphs: Given a hypergraph H with vertex set $V(H)$ and edge set $E(H)$, determine if we can 2-colour the vertices $V(H)$ such that every hyperedge in $E(H)$ contains vertices of both colours.
Set Cover Packing: The input consists of a bipartite graph $G = (C \cup B, E)$, and the goal is to find a largest collection of pairwise disjoint set covers of B, where a *set cover* of B is a subset $S \subseteq C$ such that each vertex of B has a neighbour in C.

1.2 Backgrounds

1.2.1 Steiner Tree Packing Problem in Undirected Graphs

For a graph $G = (V, E)$ and a (terminal) set $S \subseteq V$ of at least two vertices, an *S-Steiner tree* or, simply, an *S-tree* is a tree T of G with $S \subseteq V(T)$. Two S-trees T_1 and T_2 are said to be *edge-disjoint* if $E(T_1) \cap E(T_2) = \emptyset$. Two edge-disjoint S-trees T_1 and T_2 are said to be *internally-disjoint* if $V(T_1) \cap V(T_2) = S$. The basic problem of Steiner Tree Packing is defined as follows:

Steiner Tree Packing (STP): The input consists of an undirected graph G and a subset of vertices $S \subseteq V(G)$, the goal is to find a largest collection of pairwise edge-disjoint S-Steiner trees.

From a theoretical perspective, both extremes of this problem are highly related fundamental theorems in combinatorics. One extreme of the problem is when $|S| = 2$, in this case, finding edge-disjoint Steiner trees is equivalent to finding edge-disjoint paths between the two terminals in S, and so the problem is related to the well-known Menger's Theorem. The other extreme is when $|S| = n$, in this case, edge-disjoint Steiner trees are just edge-disjoint spanning trees of G, and so the problem is related to the classical Nash-Williams–Tutte Theorem [56, 81].

From a practical perspective, the Steiner tree packing problem has applications in VLSI circuit design [34, 63]. In this application, a Steiner tree is needed to share an electronic signal by a set of terminal nodes. Another application arises in the Internet Domain [53]: Let a given graph G represent a network, we choose arbitrary k vertices as nodes such that one of them is a *broadcaster*, and all other nodes are either *users* or *routers* (also called *switches*). The broadcaster wants to broadcast as many streams of movies as possible, so that the users have the maximum number of choices. Each stream of movie is broadcasted via a tree connecting all the users and the broadcaster. Hence we need to find the maximum number Steiner trees connecting all the users and the broadcaster, and it is a Steiner tree packing problem.

The Steiner tree packing problem in undirected graphs has attracted much attention from researchers in the area of graph theory, combinatorial optimization and theoretical computer science, and has become a well-established research topic [20, 22, 27, 30–34, 44, 45, 47–49, 52, 53, 59, 63, 82, 84].

1.2.2 Directed Steiner Tree Packing Problem

For a digraph $D = (V(D), A(D))$, and a set $S \subseteq V(D)$ with $r \in S$ and $|S| \geq 2$, a *directed* (S, r)*-Steiner tree* or, simply, an (S, r)*-tree* is an out-tree T rooted at r with $S \subseteq V(T)$ [20]. Two (S, r)-trees are said to be *arc-disjoint* if they have no common arc. Two arc-disjoint (S, r)-trees are said to be *internally-disjoint* if the set of common vertices of them is exactly S. Let $\kappa_{S,r}(D)$ (resp. $\lambda_{S,r}(D)$) be the maximum number of pairwise internally-disjoint (resp. arc-disjoint) (S, r)-trees in D.

Cheriyan and Salavatipour [20], Sun and Yeo [73] studied the following two directed Steiner tree packing problems:

ARC-DISJOINT DIRECTED STEINER TREE PACKING (ADSTP): The input consists of a digraph D and a subset of vertices $S \subseteq V(D)$ with a root r, the goal is to find a largest collection of pairwise arc-disjoint (S, r)-trees.
INTERNALLY-DISJOINT DIRECTED STEINER TREE PACKING (IDSTP): The input consists of a digraph D and a subset of vertices $S \subseteq V(D)$ with a root r, the goal is to find a largest collection of pairwise internally-disjoint (S, r)-trees.

It is worth mentioning that the problem of directed Steiner tree packing is highly related to two important theorems in graph theory: when $|S| = 2$, that is, $S = \{r, x\} \subseteq V(D)$, then $\lambda_{S,r}(D) = \lambda_D(r, x)$ (resp. $\kappa_{S,r}(D) = \kappa_D(r, x)$), the local arc-strong (resp. vertex-strong) connectivity of r, x in D, and therefore is related to the well-known Menger's Theorem; when $|S| = n$, the problem is equivalent to finding a largest collection of pairwise arc-disjoint out-branchings rooted at r, and therefore is related to the famous Edmonds' Branching Theorem (Theorem 1.1).

1.2.3 Directed Steiner Path Packing Problem

Sun and Zhang [74] introduced the concept of directed Steiner path packing which could be seen as a restriction of the directed Steiner tree packing problem. For a digraph $D = (V(D), A(D))$, and a set $S \subseteq V(D)$ with $r \in S$ and $|S| \geq 2$, a *directed* (S, r)*-Steiner path* or, simply, an (S, r)*-path* is a directed path P started at r with $S \subseteq V(P)$. Observe that the directed Steiner path is a generalization of the directed Hamiltonian path.

Two (S, r)-paths are said to be *arc-disjoint* if they have no common arc. Two arc-disjoint (S, r)-paths are said to be *internally-disjoint* if the set of common vertices of them is exactly S. Let $\kappa^p_{S,r}(D)$ (resp. $\lambda^p_{S,r}(D)$) be the maximum number of pairwise internally-disjoint (resp. arc-disjoint) (S, r)-paths in D. The directed Steiner path packing problems can be defined as follows:

ARC-DISJOINT DIRECTED STEINER PATH PACKING (ADSPP): The input consists of a digraph D and a subset of vertices $S \subseteq V(D)$ with a root r, the goal is to find a largest collection of pairwise arc-disjoint (S, r)-paths.
INTERNALLY-DISJOINT DIRECTED STEINER PATH PACKING (IDSPP): The input consists of a digraph D and a subset of vertices $S \subseteq V(D)$ with a root r, the goal is to find a largest collection of pairwise internally-disjoint (S, r)-paths.

1.2.4 Strong Arc Decomposition Problem

The following famous Edmonds' Branching Theorem is a fundamental theorem in the area of digraph packing theory.

Theorem 1.1 (Edmonds' Branching Theorem [24]) *A digraph $D = (V, A)$ with a special vertex r has k pairwise arc-disjoint out-branchings rooted at r if and only if there are k arc-disjoint (r, v)-paths in D for every $v \in V \setminus \{r\}$.*

Furthermore, there exists a polynomial algorithm for finding k pairwise arc-disjoint out-branchings from a given root r if they exist. However, if we ask for the existence of a pair of arc-disjoint branchings B^+_r, B^-_r such that the first is an out-branching rooted at r and the latter is an in-branching rooted at r, then the problem becomes NP-complete (see Section 9.6 of [5]). In connection with this problem, Thomassen posed the following conjecture:

Conjecture 1.1 ([79]) There exists an integer N so that every N-arc-strong digraph D contains a pair of arc-disjoint in- and out-branchings.

Observe that every strong digraph D has an out- and in-branching rooted at any vertex of D. Bang-Jensen and Yeo generalized Conjecture 1.1 as follows.

Conjecture 1.2 ([12]) There exists an integer N so that every N-arc-strong digraph D has a strong arc decomposition.

For a general digraph D, it is a hard problem to decide whether D has a strong arc decomposition.

Theorem 1.2 ([12]) *It is NP-complete to decide whether a digraph has a strong arc decomposition.*

1.2.5 Strong Subgraph Packing Problem

There is another way to extend the Steiner tree packing problem to directed graphs, note that an S-Steiner tree is a connected subgraph of G containing S. In fact, in the definition of Steiner tree packing problem, we could replace "an S-Steiner tree" by "a connected subgraph of G containing S". Therefore, we define the strong subgraph packing problem by replacing "connected" with "strongly connected" (or, simply, "strong") as follows. Let $D = (V(D), A(D))$ be a digraph of order n, S a k-subset of $V(D)$ and $2 \leq k \leq n$. A strong subgraph H of D is called an *S-strong subgraph* if $S \subseteq V(H)$. Two S-strong subgraphs are said to be *arc-disjoint* if they have no common arc. Furthermore, two arc-disjoint S-strong subgraphs are said *internally-disjoint* if the set of common vertices of them is exactly S. We use $\kappa_S(D)$ (resp. $\lambda_S(D)$) to denote the maximum number of pairwise internally-disjoint (resp. arc-disjoint) S-strong subgraphs in D [68, 76].

Sun et al. [77] defined the following two types of strong subgraph packing problems in digraphs which are analogs of the directed Steiner tree packing problem and are extensions of the undirected Steiner tree packing problem:

ARC-DISJOINT STRONG SUBGRAPH PACKING (ASSP): The input consists of a digraph D and a subset of vertices $S \subseteq V(D)$, the goal is to find a largest collection of pairwise arc-disjoint S-strong subgraphs.
INTERNALLY-DISJOINT STRONG SUBGRAPH PACKING (ISSP): The input consists of a digraph D and a subset of vertices $S \subseteq V(D)$, the goal is to find a largest collection of pairwise internally-disjoint S-strong subgraphs.

Observe that a digraph D has a strong arc decomposition if and only if $\lambda_{V(D)}(D) \geq 2$ (or, $\kappa_{V(D)}(D) \geq 2$). Therefore, the problem of ASSP (or ISSP) could be seen as an extension of the strong arc decomposition problem.

1.2.6 Directed Steiner Cycle Packing Problem

Let $D = (V(D), A(D))$ be a digraph of order n, S a k-subset of $V(D)$ and $2 \leq k \leq n$. A directed cycle C of D is called a *directed S-Steiner cycle* or, simply, an *S-cycle* if $S \subseteq V(C)$. It is worth noting that Steiner cycles have applications in the optimal design of reliable telecommunication and transportation networks [64]. Two S-cycles are said to be *arc-disjoint* if they have no common arc. Furthermore, two arc-disjoint S-cycles are *internally-disjoint* if the set of common vertices of them is exactly S. We use $\kappa_S^c(D)$ (resp. $\lambda_S^c(D)$) to denote the maximum number of pairwise internally-disjoint (resp. arc-disjoint) S-cycles in D [72].

Sun and Jin [72] defined the following problems of packing directed Steiner cycles:

ARC-DISJOINT DIRECTED STEINER CYCLE PACKING (ADSCP): The input consists of a digraph D and a subset of vertices $S \subseteq V(D)$, the goal is to find a largest collection of pairwise arc-disjoint S-cycles.

INTERNALLY-DISJOINT DIRECTED STEINER CYCLE PACKING (IDSCP): The input consists of a digraph D and a subset of vertices $S \subseteq V(D)$, the goal is to find a largest collection of pairwise internally-disjoint S-cycles.

By definition, the directed Steiner cycle packing problem is a special restriction of the strong subgraph packing problem, as an S-cycle is also an S-strong subgraph. However, the directed Steiner cycle packing problem is quite distinct from strong subgraph packing problem. For example, observe that it can be decided in polynomial-time whether $\kappa_S(D) \geq 1$, but it is NP-complete to decide whether $\kappa_S^c(D) \geq 1$ even restricted to an Eulerian digraph D (see Theorem 6.1).

The directed Steiner cycle packing problem is also related to other problems in graph theory. When $|S| = n$, an S-cycle is a directed Hamiltonian cycle. Therefore, the directed Steiner cycle packing problem generalizes the directed Hamiltonian cycle packing problem (and therefore the directed Hamiltonian decomposition problem). A digraph D is *k-cyclic* if it has a cycle containing the vertices $x_1, x_2, \ldots, x_k$ for every choice of k vertices. Note that the notion of k-cyclic attracts the attention of some researchers, such as [50]. By definition, a digraph is k-cyclic if and only if $\kappa_S^c(D) \geq 1$ (resp. $\lambda_S^c(D) \geq 1$) for every k-subset S of $V(D)$.

1.2.7 Steiner Type Packing Problems in Digraphs

Generally, for a digraph D and a subset of vertices $S \subseteq V(D)$, a *directed S-Steiner type subgraph* is a subgraph H of D such that $S \subseteq V(H)$. Two directed S-Steiner type subgraphs are said to be *arc-disjoint* if they have no common arc. Furthermore, two arc-disjoint directed S-Steiner type subgraphs are said *internally-disjoint* if the set of common vertices of them is exactly S. We now define the following problems:

ARC-DISJOINT DIRECTED STEINER TYPE SUBGRAPH PACKING (ADSTSP): The input consists of a digraph D and a subset of vertices $S \subseteq V(D)$, the goal is to find a largest collection of pairwise arc-disjoint directed S-Steiner type subgraphs.

INTERNALLY-DISJOINT DIRECTED STEINER TYPE SUBGRAPH PACKING (IDSTSP): The input consists of a digraph D and a subset of vertices $S \subseteq V(D)$, the goal is to find a largest collection of pairwise internally-disjoint directed S-Steiner type subgraphs.

ADSTSP and IDSTSP are also called *Steiner type packing problems*. By definition, all of ADSTP, IDSTP, ADSPP, IDSPP, ASSP, ISSP (and therefore strong arc decomposition problem), ADSCP and IDSCP belong to this type of problem.

Chapter 2
Directed Steiner Tree Packing Problem

Abstract In the two former sections, we introduce the complexity for the decision versions of IDSTP and ADSTP on general digraphs, symmetric digraphs and Eulerian digraphs. The hardness of approximation of ADSTP and IDSTP and their generalizations are also mentioned in Sect. 2.1. In the last section, Sect. 2.3, we introduce three related topics: directed tree connectivity, independent directed Steiner trees and independent branchings, and arc-disjoint in- and out-branchings rooted at the same vertex.

2.1 General Digraphs

Fortune, Hopcroft and Wyllie proved the following important theorem on DIRECTED 2-LINKAGE.

Theorem 2.1 ([26]) *The* DIRECTED 2-LINKAGE *is NP-complete.*

Let D be a digraph and $S \subseteq V(D)$ with $r \in S$ and $|S| = k$. It is natural to consider the following decision version of IDSTP (resp. ADSTP): what is the complexity of deciding whether $\kappa_{S,r}(D) \geq \ell$ (resp. $\lambda_{S,r}(D) \geq \ell$)? If $k = 2$, say $S = \{r, x\}$, then the problem of deciding whether $\kappa_{S,r}(D) \geq \ell$ (resp. $\lambda_{S,r}(D) \geq \ell$) is equivalent to deciding whether $\kappa(r, x) \geq \ell$ (resp. $\lambda(r, x) \geq \ell$), and so is polynomial-time solvable (see [5]), where $\kappa(r, x)$ (resp. $\lambda(r, x)$) is the local vertex-strong (resp. arc-strong) connectivity from r to x. If $\ell = 1$, then the above problem is also polynomial-time solvable by the well-known fact that every strong digraph has an out- and in-branching rooted at any vertex, and these branchings can be found in polynomial-time. Hence, it remains to consider the case that $k \geq 3, \ell \geq 2$. Using Theorem 2.1, Cheriyan and Salavatipour proved the NP-hardness of the case $k = 3, \ell = 2$ for both $\kappa_{S,r}(D)$ and $\lambda_{S,r}(D)$ (Theorems 2.2 and 2.3). We just give the proof of Theorem 2.2, Theorem 2.3 can be deduced from Theorem 2.2 and we omit the details.

Theorem 2.2 ([20]) *Let D be a digraph and $S \subseteq V(D)$ with $|S| = 3$. The problem of deciding whether $\kappa_{S,r}(D) \geq 2$ is NP-hard, where $r \in S$.*

Y. Sun, *Steiner Type Packing Problems in Digraphs*, SpringerBriefs in Mathematics,
https://doi.org/10.1007/978-981-95-8743-8_2

Proof Let $I = [D; x_1, y_1, x_2, y_2]$ be an instance of DIRECTED 2-LINKAGE problem. Construct D' from D by adding three new vertices r, t_1, t_2 and six new arcs $rx_1, rx_2, y_1t_1, x_2t_1, y_2t_2, x_1t_2$. Let $S = \{r, t_1, t_2\}$ and let r be the root. We claim that I is a positive instance of DIRECTED 2-LINKAGE problem if and only if $\kappa_{S,r}(D') \geq 2$. If I is a positive instance of DIRECTED 2-LINKAGE problem, that is, there are two disjoint paths P_1 and P_2 such that P_i is from x_i to y_i for $i \in [2]$. Then $P_1 \cup \{rx_1, y_1t_1, x_1t_2\}$ and $P_2 \cup \{rx_2, x_2t_1, y_2t_2\}$ form two internally-disjoint (S, r)-trees in D', which means that $\kappa_{S,r}(D') \geq 2$. Conversely, let T_1 and T_2 be two internally-disjoint (S, r)-trees in D'. Since r has only two out-arcs, we may assume that $rx_i \in A(T_i)$ for $i \in [2]$. Therefore, there is a path from x_1 to t_1 in T_1, which must go through y_1 (since x_2 is not in T_1), and a path from x_2 to t_2 in T_2, which must go through y_2 (since x_1 is not in T_2). These two paths are disjoint as T_1 and T_2 are internally-disjoint. Hence, I is a positive instance of DIRECTED 2-LINKAGE problem. Now by Theorem 2.1, the result holds. □

Theorem 2.3 ([20]) *Let D be a digraph and $S \subseteq V(D)$ with $|S| = 3$. The problem of deciding whether $\lambda_{S,r}(D) \geq 2$ is NP-hard, where $r \in S$.*

Sun and Yeo extended Theorems 2.2 and 2.3 to the following.

Theorem 2.4 ([73]) *Let $k \geq 3$ and $\ell \geq 2$ be fixed integers (considered as constants). Let D be a digraph and $S \subseteq V(D)$ with $|S| = k$ and $r \in S$. Both the following problems are NP-complete.*

- *Is $\kappa_{S,r}(D) \geq \ell$?*
- *Is $\lambda_{S,r}(D) \geq \ell$?*

Proof We first consider the problem of determining if $\kappa_{S,r}(D) \geq \ell$. It is easy to see that this problem belongs to NP. To show it is NP-hard, we reduce from the case when $k = 3$ and $\ell = 2$. That is let D^* be a digraph such that $S^* \subseteq V(D^*)$ with $|S^*| = 3$ and $r \in S^*$ where we want to determine if $\kappa_{S^*,r}(D^*) \geq 2$. This problem is NP-hard by Theorem 2.2.

We will construct a new digraph D containing a set $S \subseteq V(D)$ with $|S| = k$ and $r \in S$, such that $\kappa_{S,r}(D) \geq \ell$ if and only if $\kappa_{S^*,r}(D^*) \geq 2$. This would complete the proof for the problem of determining if $\kappa_{S,r}(D) \geq \ell$. Assume that $S^* = \{r, s_1, s_2\}$ and let $V(D) = V(D^*) \cup U \cup W \cup \{s_3, s_4, \ldots, s_{k-1}\}$, where $U = \{u_1, u_2, \ldots, u_{\ell-2}\}$ and $W = \{w_1, w_2, \ldots, w_{\ell-2}\}$.

Let $S = \{r, s_1, s_2, s_3, \ldots, s_{k-1}\}$ and let the arc-set of D be the following (see Fig. 2.1).

$$\begin{aligned} A(D) = A(D^*) &\cup \{ru_i, u_iw_i, w_is \mid i \in [\ell - 2] \text{ and } s \in S - r\} \\ &\cup \{s_is_j \mid i \in [2] \text{ and } j = 3, 4, \ldots, k - 1\} \end{aligned}$$

First assume that $\kappa_{S^*,r}(D^*) \geq 2$ and let T_1^* and T_2^* be the two internally-disjoint (S^*, r)-trees in D^*. Add all arcs from s_1 to $\{s_3, s_4, \ldots, s_{k-1}\}$ to T_1^* and add all arcs from s_2 to $\{s_3, s_4, \ldots, s_{k-1}\}$ to T_2^* in order to obtain two internally-disjoint (S, r)-trees in D. Furthermore, for $i \in [\ell - 2]$, we note that the path ru_iw_i together with

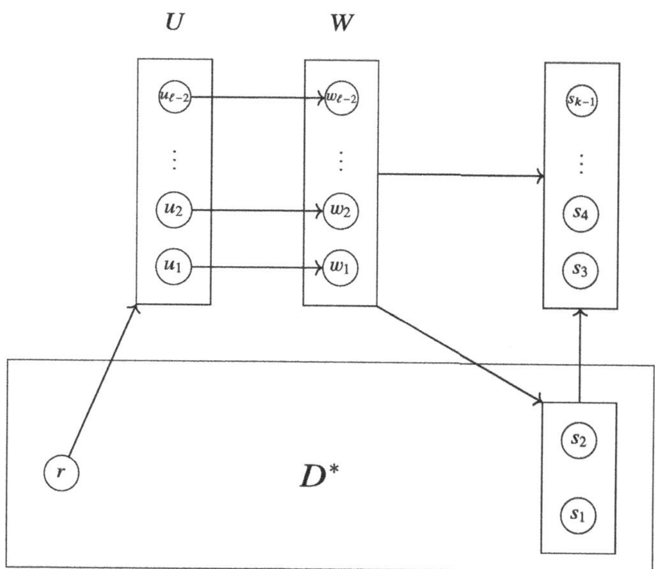

Fig. 2.1 In the above digraph, D, the following statements hold. (1): $\kappa_{\{r,s_1,s_2,\ldots,s_{k-1}\},r}(D) \geq \ell$ if and only if $\kappa_{\{r,s_1,s_2\},r}(D^*) \geq 2$. (2): $\lambda_{\{r,s_1,s_2,\ldots,s_{k-1}\},r}(D) \geq \ell$ if and only if $\lambda_{\{r,s_1,s_2\},r}(D^*) \geq 2$

Table 2.1 Directed graphs

$\lambda_{S,r}(D) \geq \ell$? $\lvert S\rvert = k$	$k = 3$	$k \geq 4$ constant	k part of input
$\ell = 2$	NP-complete [20]	NP-complete [73]	NP-complete [73]
$\ell \geq 3$ constant	NP-complete [73]	NP-complete [73]	NP-complete [73]
ℓ part of input	NP-complete [73]	NP-complete [73]	NP-complete [73]

all arcs from w_i to $S \setminus \{r\}$ gives us an (S, r)-tree in D. It is not difficult to see that the ℓ constructed (S, r)-trees are all internally-disjoint and therefore $\kappa_{S,r}(D) \geq \ell$.

Conversely, if $\kappa_{S,r}(D) \geq \ell$, then let $T_1, T_2, \ldots, T_\ell$ be the ℓ internally-disjoint (S, r)-trees in D. If we remove all the T_i's containing a vertex from U then we are left with at least two trees, not containing any vertex from U or W (as after removing U the vertices in W have no arc into them). Removing all vertices from $\{s_3, s_4, \ldots, s_{k-1}\}$ from the two remaining T_i's gives us two internally-disjoint (S^*, r)-trees in D^* and therefore $\kappa_{S^*,r}(D^*) \geq 2$.

This completes the proof for determining if $\kappa_{S,r}(D) \geq \ell$. The case when we want to determine if $\lambda_{S,r}(D) \geq \ell$ is similar. We use exactly the same reduction, except in D^* we look at the problem of determining if $\lambda_{S^*,r}(D^*) \geq 2$, which is also NP-hard by Theorem 2.3. Otherwise the reduction is identical (but delete the arcs $u_i w_i$ instead of the vertices in U in the above proof). □

Theorems 2.2, 2.3 and 2.4 imply the entries in Tables 2.1 and 2.2.

Table 2.2 Directed graphs

$\kappa_{S,r}(D) \geq \ell$? $\lvert S\rvert = k$	$k = 3$	$k \geq 4$ constant	k part of input
$\ell = 2$	NP-complete [20]	NP-complete [73]	NP-complete [73]
$\ell \geq 3$ constant	NP-complete [73]	NP-complete [73]	NP-complete [73]
ℓ part of input	NP-complete [73]	NP-complete [73]	NP-complete [73]

For general digraphs, Cheriyan and Salavatipour [20] also studied the hardness of approximation of both ADSTP and IDSTP. They proved that given an instance of ADSTP, it is NP-hard to approximate the solution within $O(m^{1/3-\epsilon})$ for any $\epsilon > 0$. Similarly, given an instance of IDSTP, it is NP-hard to approximate the solution within $O(n^{1/3-\epsilon})$ for any $\epsilon > 0$.

There are two more general problems:
GDE: The input consists of a digraph D, a capacity c_e on each arc $e \in A(D)$, ℓ terminal sets $T_1, \ldots, T_\ell$ and ℓ roots $r_1, \ldots, r_\ell$ such that $r_i \in V(T_i)$ for each $i \in [\ell]$. The goal is to find a largest collection of directed Steiner trees, each rooted at an r_i and containing all vertices of T_i such that each arc e is contained in at most c_e directed trees.
GDV: The input consists of a digraph D, a capacity c_v on each vertex $v \in V(D)$, ℓ terminal sets $T_1, \ldots, T_\ell$ and ℓ roots $r_1, \ldots, r_\ell$ such that $r_i \in V(T_i)$ for each $i \in [\ell]$. The goal is to find a largest collection of directed Steiner trees, each rooted at an r_i and containing all vertices of T_i such that each non-terminal vertex v is contained in at most c_v directed trees.

Clearly, when $\ell = 1$ and $c_e = 1$ (resp. $c_v = 1$), then GDE (resp. GDV) is exactly ADSTP (resp. IDSTP). It is worth mentioning that the hardness of approximation of both GDE and GDV has also been studied by researchers [20, 36].

2.2 Symmetric Digraphs and Eulerian Digraphs

Now we turn our attention to symmetric digraphs. We first need the following theorem by Robertson and Seymour.

Theorem 2.5 ([60]) *Let $k \geq 2$ be a fixed integer. Let G be a graph and let $s_1, s_2, \ldots, s_k, t_1, t_2, \ldots, t_k$ be $2k$ disjoint vertices in G. We can in $O(|V(G)|^3)$ time decide if there exists an (s_i, t_i)-path, P_i, such that all $P_1, P_2, \ldots, P_k$ are vertex disjoint.*

Using Theorem 2.5, Sun and Yeo proved the following result which implies the problem of DIRECTED k-LINKAGE can be solved in polynomial-time for symmetric digraphs, when k is fixed.

Corollary 2.1 ([73]) *Let D be a symmetric digraph and let $s_1, s_2, \ldots, s_k, t_1, t_2, \ldots, t_k$ be vertices in D (not necessarily disjoint) and let $S \subseteq V(D)$. We can* *in $O(|V(D)|^3)$ time decide if there for all $i \in [k]$ exists an (s_i, t_i)-path, P_i, such that no internal vertex of any P_i belongs to S or to any path P_j with $j \neq i$ (the end-points of P_j can also not be internal vertices of P_i).*

Furthermore, Sun and Yeo proved the following result on symmetric digraphs by Corollary 2.1.

Theorem 2.6 ([73]) *Let $k \geq 3$ and $\ell \geq 2$ be fixed integers and let D be a symmetric digraph. Let $S \subseteq V(D)$ with $|S| = k$ and let r be an arbitrary vertex in S. Let $A_0, A_1, A_2, \ldots, A_\ell$ be a partition of the arcs in $D[S]$.*

We can in time $O(n^{\ell k-2\ell+3} \cdot (2k-3)^{\ell(2k-3)})$ decide if there exist ℓ pairwise internally-disjoint (S, r)-trees, $T_1, T_2, \ldots, T_\ell$, with $A(T_i) \cap A(D[S]) = A_i$ for all $i \in [\ell]$ (note that A_0 are the arcs in $D[S]$ not used in any of the trees).

Proof Let T be any (S, r)-tree in D. Let R be all vertices in T with degree (in the underlying graph of T) at least three and let L be all the leaves of T (in the underlying graph of T). It is well-known that $|R| \leq |L| - 2$. Furthermore we will assume that all (S, r)-trees considered are minimal (i.e. if we delete a vertex in the tree it is not an (S, r)-tree anymore), which implies that all leaves are in S. Under this assumption we have $|R| \leq |L| - 2 \leq |S| - 2$. Note that every vertex $x \in V(T) \setminus (R \cup S)$ has $d_T^+(x) = d_T^-(x) = 1$. We now define the *skeleton* of T as the tree we obtain from T by by-passing all vertices in $V(T) \setminus (R \cup S)$ (i.e. if $ux, xv \in A(T)$ and $x \in V(T) \setminus (R \cup S)$ then replace ux and xv by uv).

Let T^s be a skeleton of an (S, r)-tree in D and recall that $V(T^s)$ consists of S as well as at most $|S| - 2$ $(= k - 2)$ other vertices. Therefore there are less than n^{k-2} possibilities for $V(T^s)$. Once $V(T^s)$ is known there are at most $(2k-3)^{2k-3}$ possibilities for the arcs of T^s (as for each $y \in V(T^s) \setminus \{r\}$ we need to pick the vertex with an arc into y and there are at most $2k - 3$ possibilities for picking this vertex). This implies that there are at most $n^{k-2} \cdot (2k-3)^{2k-3}$ different skeletons of minimal (S, r)-trees in D (as there are at most n^{k-2} possibilities for $V(T^s)$ and at most $(2k-3)^{2k-3}$ possibilities for $E(T^s)$ for each given $V(T^s)$).

Our algorithm will try all possible ℓ-tuples, $\mathcal{T}^s = (T_1^s, T_2^s, \ldots, T_\ell^s)$, of skeletons of (S, r)-trees and determine if there are ℓ internally-disjoint (S, r)-trees, $T_1, T_2, \ldots, T_\ell$, with $A(T_i) \cap A(D[S]) = A_i$ for all $i \in [\ell]$ and such that T_i^s is the skeleton of T_i. If such a set of trees exists for any $\mathcal{T}^s$, then we return this solution and if no such set of trees exists for any $\mathcal{T}^s$ we return that no solution exists. We will first prove that this algorithm gives the correct answer and then compute its time complexity.

If our algorithm returns a solution, then clearly a solution exists. So now assume that a solution exists and let $T_1, T_2, \ldots, T_\ell$ be the desired set of internally-disjoint (S, r)-trees. When we consider $\mathcal{T}^s = (T_1^s, T_2^s, \ldots, T_\ell^s)$, where T_i^s is the skeleton of T_i, our algorithm will find a solution, so the algorithm always returns a solution if one exists.

We will now analyse the time complexity. The number of different ℓ-tuples, $\mathcal{T}^s$, that we need to consider is bounded by the following.

$$\left(n^{k-2} \cdot (2k-3)^{2k-3}\right)^{\ell} = n^{\ell(k-2)} \cdot (2k-3)^{\ell(2k-3)}$$

Given such a ℓ-tuples, $\mathcal{T}^s$, we need to determine if there exist ℓ internally-disjoint (S, r)-trees, $T_1, T_2, \ldots, T_\ell$, with $A(T_i) \cap A(D[S]) = A_i$ for all $i \in [\ell]$ and such that T_i^s is the skeleton of T_i. We first check that the arcs in A_i belong to the skeleton T_i^s and that no vertex in $V(D) \setminus S$ belongs to more than one skeleton. If the above does not hold then the desired trees do not exist, so assume the above holds. We will now use Corollary 2.1. In fact, for every arc $uv \notin A(D[S])$ that belongs to some skeleton T_i^s we want to find a (u, v)-path in $D - A(D[S])$, such that no internal vertex on any path belongs to S or to a different path. By Corollary 2.1 this can be done in $O(n^3)$ time. If such paths exist, then substituting the arcs uv by the (u, v)-paths we obtain the desired (S, r)-trees. And if the paths do not exist the desired (S, r)-trees do not exist. Therefore the algorithm works correctly and has complexity $O(n^{\ell k-2\ell+3} \cdot (2k-3)^{\ell(2k-3)})$. □

The next corollary implies all the polynomial entries in Table 2.4.

Corollary 2.2 ([73]) *Let $k \geq 3$ and $\ell \geq 2$ be fixed integers. We can in polynomial time decide if $\kappa_{S,r}(D) \geq \ell$ for any symmetric digraph, D, with $S \subseteq V(D)$, with $|S| = k$ and $r \in S$.*

Proof Let D be any symmetric digraph with $S \subseteq V(D)$, with $|S| = k$ and $r \in S$. Let $\mathcal{A} = (A_0, A_1, A_2, \ldots, A_\ell)$ be a partition of the arcs in $D[S]$. By Theorem 2.6 we can decide if there exist ℓ internally-disjoint (S, r)-trees, $T_1, T_2, \ldots, T_\ell$, with $A(T_i) \cap A(D[S]) = A_i$ for all $i \in [\ell]$. If such ℓ internally-disjoint (S, r)-trees exist then clearly $\kappa_{S,r}(D) \geq \ell$. We will do the above for all possible partitions $\mathcal{A} = (A_0, A_1, A_2, \ldots, A_\ell)$ and if we find ℓ internally-disjoint (S, r)-trees for any such partition then we return "$\kappa_{S,r}(D) \geq \ell$" and otherwise we return "$\kappa_{S,r}(D) < \ell$".

If $\kappa_{S,r}(D) \geq \ell$ then we note that we will correctly determine that $\kappa_{S,r}(D) \geq \ell$, when we consider the correct partition $\mathcal{A}$, which proves that the above algorithms will always return the correct answer.

Furthermore since the number of partitions, $\mathcal{A}$, of $A(D[S])$ is bounded by $(\ell + 1)^{|A(D[S])|} \leq (\ell+1)^{k^2/2}$, we note that the above algorithm runs in polynomial time (as ℓ and k are considered constants). □

We now turn our attention to the NP-complete cases in Table 2.4. Chen et al. [19] introduced the CLLM PROBLEM, which turned out to be NP-complete.

Lemma 2.1 ([19]) *The* CLLM PROBLEM *is NP-complete.*

Using Lemma 2.1, Sun and Yeo showed the following.

Theorem 2.7 ([73]) *Let $k \geq 3$ be a fixed integer. The problem of deciding if a symmetric digraph D, with a k-subset S of $V(D)$ with $r \in S$ satisfies $\kappa_{S,r}(D) \geq \ell$ (ℓ is part of the input), is NP-complete.*

Proof It is easy to see that this problem is in NP. Let G be a tripartite graph with 3-partition (A, B, C) such that $|A| = |B| = |C| = \ell$. We will construct a symmetric digraph D, a k-subset $S \subseteq V(D)$ with $r \in S$ and an integer ℓ such that there are ℓ internally-disjoint (S, r)-trees in D if and only if G is a positive instance of the CLLM PROBLEM.

Let D be obtained from G by replacing every edge with a 2-cycle and adding the vertices $S = \{r, s_1, s_2, \ldots, s_{k-1}\}$ and all arcs between r and A and all arcs between s_1 and B and all arcs between $\{s_2, s_3, \ldots, s_{k-1}\}$ and C. This completes the construction of D. Note that the construction of D clearly can be done in polynomial time.

First consider the case when there are ℓ internally-disjoint (S, r)-trees in D, say T_i $(i \in [\ell])$. Each tree must clearly contain at least one vertex from A (connected to r in the tree), at least one vertex from B (connected to s_1 in the tree) and at least one vertex from C (connected to s_2 in the tree). As $|A| = |B| = |C| = \ell$ and the trees are internally-disjoint, we note that every tree, T_i, contains exactly one vertex from A, say a_i, and one vertex from B, say b_i, and exactly one vertex from C, say c_i. However now $G[\{a_i, b_i, c_i\}]$ is connected for all $i \in [\ell]$, and G is a positive instance of CLLM PROBLEM.

Conversely if G is a positive instance of CLLM PROBLEM, then there is a partition of $V(G)$ into ℓ disjoint sets $V_1, V_2, \ldots, V_\ell$ each having three vertices, such that for every $V_i = \{a_{i_1}, b_{i_2}, c_{i_3}\}$ we have $a_{i_1} \in A$, $b_{i_2} \in B$ and $c_{i_3} \in C$, and $G[V_i]$ is connected. Let T_i be an (S, r)-tree with vertex set $V_i \cup S$. Note that all T_i are internally-disjoint (S, r)-trees in D. By the above argument and Lemma 2.1, we are done. □

The problem of 2-COLORING HYPERGRAPHS is known to be NP-hard by Lovász.

Theorem 2.8 ([55]) *The problem of* 2-COLORING HYPERGRAPHS *is NP-hard.*

By constructing a reduction from the problem of 2-COLORING HYPERGRAPHS, Sun and Yeo proved the following theorem.

Theorem 2.9 ([73]) *Let $\ell \geq 2$ be a fixed integer. The problem of deciding if a symmetric digraph D, with an $S \subseteq V(D)$ and $r \in S$ satisfies $\kappa_{S,r}(D) \geq \ell$ ($k = |S|$ is part of the input), is NP-complete.*

Proof We will reduce from the problem of 2-COLORING HYPERGRAPHS (see [55]). That is, we are given a hypergraph, H, with vertex set $V(H)$ and edge set $E(H)$, and want to determine if we can 2-colour the vertices $V(H)$ such that every hyperedge in $E(H)$ contains vertices of both colours. This problem is known to be NP-hard by Theorem 2.8.

Define a symmetric digraph, D, as follows. Let $U = \{u_1, u_2, \ldots, u_{\ell-2}\}$ and let $V(D) = V(H) \cup E(H) \cup U \cup \{r\}$ and let the arc-set of D be defined as follows.

$$\begin{aligned} A(D) = \{&xe, ex \mid x \in V(H),\ e \in E(H) \text{ and } x \in V(e)\} \\ &\cup\ \{ru_i, u_ir, u_ie, eu_i \mid u_i \in U \text{ and } e \in E(H)\} \\ &\cup\ \{rx, xr \mid x \in V(H)\} \end{aligned}$$

Table 2.3 Symmetric digraphs

$\lambda_{S,r}(D) \geq \ell$? $\lvert S\rvert = k$	$k = 3$	$k \geq 4$ constant	k part of input
$\ell = 2$	Polynomial [73]	Polynomial [73]	Polynomial [73]
$\ell \geq 3$ constant	Polynomial [73]	Polynomial [73]	Polynomial [73]
ℓ part of input	Polynomial [73]	Polynomial [73]	Polynomial [73]

Table 2.4 Symmetric digraphs

$\kappa_{S,r}(D) \geq \ell$? $\lvert S\rvert = k$	$k = 3$	$k \geq 4$ constant	k part of input
$\ell = 2$	Polynomial [73]	Polynomial [73]	NP-complete [73]
$\ell \geq 3$ constant	Polynomial [73]	Polynomial [73]	NP-complete [73]
ℓ part of input	NP-complete [73]	NP-complete [73]	NP-complete [73]

Let $S = E(H) \cup \{r\}$. This completes the construction of D, S and r. We will show that $\kappa_{S,r}(D) \geq \ell$ if and only if H is 2-colourable, which will complete the proof.

First assume that H is 2-colourable and let R be the red vertices in H and B be the blue vertices in H in a proper 2-colouring of H. Let T_i contain the arc ru_i and all arcs from u_i to $S \setminus \{r\}$ for $i \in [\ell - 2]$. Let $T_{\ell-1}$ contain all arcs from r to R and for each edge $e \in E(H)$ we add an arc from R to e in D to $T_{\ell-1}$ (this is possible as every edge in H contains a red vertex). Analogously, let T_ℓ contain all arcs from r to B and for each edge $e \in E(H)$ we add an arc from B to e in D to T_ℓ (again, this is possible as every edge in H contains a blue vertex). We now note that $T_1, T_2, \ldots, T_\ell$ are internally-disjoint (S, r)-trees in D, so $\kappa_{S,r}(D) \geq \ell$.

Conversely assume that $\kappa_{S,r}(D) \geq \ell$ and let $T'_1, T'_2, \ldots, T'_\ell$ be ℓ internally-disjoint (S, r)-trees in D. At least two of these trees contain no vertex from U (as $|U| = \ell - 2$). Without loss of generality assume that T'_1 and T'_2 don't contain any vertex from U. Let B' be all out-neighbours of r in T'_1 (i.e. $B' = N^+_{T'_1}(r)$) and let R' be all out-neighbours of r in T'_2 (i.e. $R' = N^+_{T'_2}(r)$). For every $e \in E(H)$ it has an arc into it in T'_1 and an arc into it in T'_2 which implies that in H the edge e contains a vertex from B' and a vertex from R'. Therefore H is 2-colourable (any vertex in H that is not in either R' or B' can be assigned randomly to either B' or R'). This completes the proof. □

Theorem 2.7 together with Theorem 2.9 implies all the NP-completeness results in Table 2.4.

In a digraph D, the local arc-strong connectivity is the maximum number of arc-disjoint (x, y)-paths and is denoted by $\lambda_D(x, y)$, or just by $\lambda(x, y)$ if D is clear from the context.

Theorem 2.10 ([13]) *Let $k \geq 1$ and let $D = (V, A)$ be a directed multigraph with a special vertex r. Let $T' = \{x \mid x \in V \setminus \{r\}$ and $d^-(x) < d^+(x)\}$. If $\lambda(r, x) \geq k$*

Table 2.5 Eulerian digraphs

$\lambda_{S,r}(D) \geq \ell$? $\|S\| = k$	$k = 3$	$k \geq 4$ constant	k part of input
$\ell = 2$	Polynomial [73]	Polynomial [73]	Polynomial [73]
$\ell \geq 3$ constant	Polynomial [73]	Polynomial [73]	Polynomial [73]
ℓ part of input	Polynomial [73]	Polynomial [73]	Polynomial [73]

for every $x \in T'$, then there exists a family $\mathcal{F}$ of k arc-disjoint out-trees rooted at r so that every vertex $x \in V$ belongs to at least $\min\{k, \lambda(r, x)\}$ *members of $\mathcal{F}$.*

In the case when D is Eulerian, then $d^+(x) = d^-(x)$ for all $x \in V(D)$ and therefore $T' = \emptyset$ in the above theorem. Therefore the following corollary holds.

Corollary 2.3 ([73]) *Let $k \geq 1$ and let $D = (V, A)$ be an Eulerian digraph with a special vertex r. Then there exists a family $\mathcal{F}$ of k arc-disjoint out-trees rooted at r so that every vertex $x \in V$ belongs to at least* $\min\{k, \lambda(r, x)\}$ *members of $\mathcal{F}$.*

Using Corollary 2.3, Sun and Yeo proved the following Theorem 2.11. As one can determine $\lambda_D(r, s)$ in polynomial time for any r and s in D, we note that Theorem 2.11 implies that Table 2.5 (and therefore Table 2.3) holds.

Theorem 2.11 ([73]) *If D is an Eulerian digraph and $S \subseteq V(D)$ and $r \in S$, then $\lambda_{S,r}(D) \geq \ell$ if and only if $\lambda_D(r, s) \geq \ell$ for all $s \in S \setminus \{r\}$.*

Proof Let D be an Eulerian digraph and let $S \subseteq V(D)$ and $r \in S$ be arbitrary. First assume that $\lambda_{S,r}(D) \geq \ell$. This implies that $\lambda_D(r, s) \geq \ell$ for all $s \in S \setminus \{r\}$ as there is a path from r to s in each of the ℓ arc-disjoint (S, r)-trees and these ℓ paths are therefore also arc-disjoint.

Now assume that $\lambda_D(r, s) \geq \ell$ for all $s \in S \setminus \{r\}$. By Corollary 2.3 there exists a family $\mathcal{F}$ of ℓ arc-disjoint out-trees rooted at r so that every vertex $x \in V$ belongs to at least $\min\{\ell, \lambda(r, x)\}$ members of $\mathcal{F}$. As $\lambda_D(r, s) \geq \ell$ for all $s \in S \setminus \{r\}$, we note that every vertex in S belongs to all ℓ out-trees in $\mathcal{F}$. Therefore $\lambda_{S,r}(D) \geq \ell$, which completes the proof of this theorem. □

In the proof of Theorem 2.6 we saw how the fact that DIRECTED k-LINKAGE in symmetric digraphs is polynomial was used to prove the polynomial entries in Table 2.4. It is maybe therefore slightly surprising that even though we prove that all entries in Table 2.5 are polynomial, this could not be proved using a similar linkage result, due to the following.

Theorem 2.12 ([43]) DIRECTED WEAK k-LINKAGE *is NP-hard in Eulerian digraphs (where k is part of the input).*

Recently, Sun and Yeo proved that Theorem 2.1 holds even when D is Eulerian. In their argument, they used the constructions from the proof for Theorem 2.1 by Fortune, Hopcroft and Wyllie. And in fact, they used Figures 10.1–10.3 in the argument of Theorem 10.2.1 of [5].

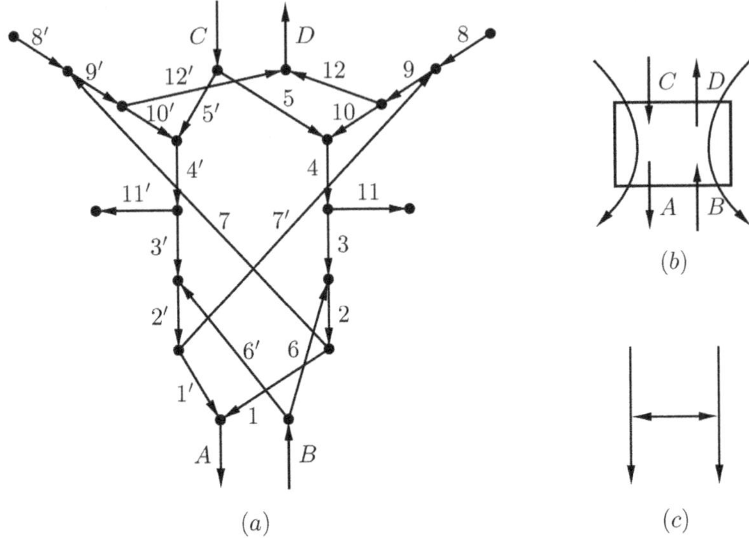

Fig. 2.2 A switch (**a**) and its schematic pictures (**b**) and (**c**) [5, 26]

In their argument, Fortune, Hopcroft and Wyllie used the concept of *switch* which is shown in Fig. 2.2a. Note that in (c) the two vertical arcs correspond to the paths $(8, 9, 10, 4, 11)$, respectively, $(8', 9', 10', 4', 11')$. For convenience, we label the arcs, rather than the vertices in this figure.

By the definition of a switch, the following lemma holds:

Lemma 2.2 ([26]) *Consider the digraph S shown in Fig. 2.2a. Suppose there are two vertex-disjoint paths P, Q passing through S such that P leaves S at A and Q enters S at B. Then P must enter S at C and Q must leave S at D. Furthermore, there exists exactly one more path R passing through S which is disjoint from P, Q and this is either* $(8, 9, 10, 4, 11)$ *or* $(8', 9', 10', 4', 11')$, *depending on the actual routing of P.*

As shown in [5, 26], we can stack arbitrarily many switches on top of each other and still have the conclusion on Lemma 2.2 holding for each switch. The way we stack is simply by identifying the *C* and *D* arcs of one switch with the *A* and *B* arcs of the next (see Fig. 2.3).

Now we can prove the NP-completeness of Directed 2-Linkage for Eulerian digraphs.

Theorem 2.13 ([73]) *The* Directed 2-Linkage *restricted to Eulerian digraphs is NP-complete.*

Proof The reduction of the argument for Theorem 2.1 is from 3-SAT problem. Let $\mathcal{F} = C_1 \wedge C_2 \wedge \cdots \wedge C_r$ be an instance of 3-SAT with variables $x_1, x_2, \ldots, x_k$ and clauses $C_1, C_2, \ldots, C_r$. For each variable x_i we let H_i be a digraph consisting of

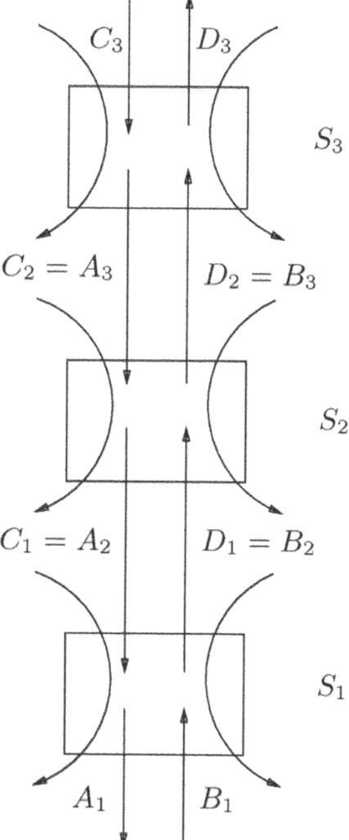

Fig. 2.3 Stacking three switches on top of each other [5, 26]

two internally-disjoint (z_i, w_i)-paths of length r (the number of clauses in $\mathcal{F}$). See Fig. 2.4 for an illustration. We associate one of these paths with the literal x_i and the other with $\overline{x}_i$.

We will now construct the digraph $D[\mathcal{F}]$ as follows (see Fig. 2.4 and e.g. [5]). Firstly, we form a chain $H_1 \rightarrow H_2 \rightarrow \cdots \rightarrow H_k$ on the subdigraphs corresponding to each variable (see the middle of the figure, H_i corresponds to the variable x_i). Secondly, with each clause C_i we associate three switches, one for each literal it contains, and they are denoted by $S_{i,1}$, $S_{i,2}$, $S_{i,3}$, respectively; we then stack these switches in the order $S_{1,1}, S_{1,2}, S_{1,3}, S_{2,1}, \dots, S_{r,1}, S_{r,2}, S_{r,3}$ as shown in the right part of the figure.

Thirdly, we create a path $n_0 n_1 \dots n_r$, such that there are three arcs from n_{i-1} to n_i for all $i \in [r]$, each corresponding to a literal in C_i. And if x_a is the b'th literal in C_i, then we substitute the b'th arc from n_{i-1} to n_i with the "left" path of $S_{i,b}$ (that is, the path with arcs $8', 9', 10', 4', 11'$ in $S_{i,b}$) and we substitute a (private) arc of H_a, such that the arc is taken from the path which corresponds to x_a if the literal is x_a and from the path which corresponds to $\overline{x}_a$ if the literal is $\overline{x}_a$, with the "right" path of $S_{i,b}$ (that is, the path with arcs 8, 9, 10, 4, 11 in $S_{i,b}$). For example in Fig. 2.4 this would imply that the right-most arc from n_{i-1} to n_i actually is the path $8', 9', 10', 4', 11'$ in $S_{i,1}$ (as x_1 is the first literal in C_i) and the arc indicated in H_1 is actually the path 8, 9, 10, 4, 11 in $S_{i,1}$. In this way one can show that only one of the paths $8', 9', 10', 4', 11'$ and 8, 9, 10, 4, 11 in each $S_{i,b}$ will ever be used in a solution to the 2-linkage problem. In other words, The double arcs which point to

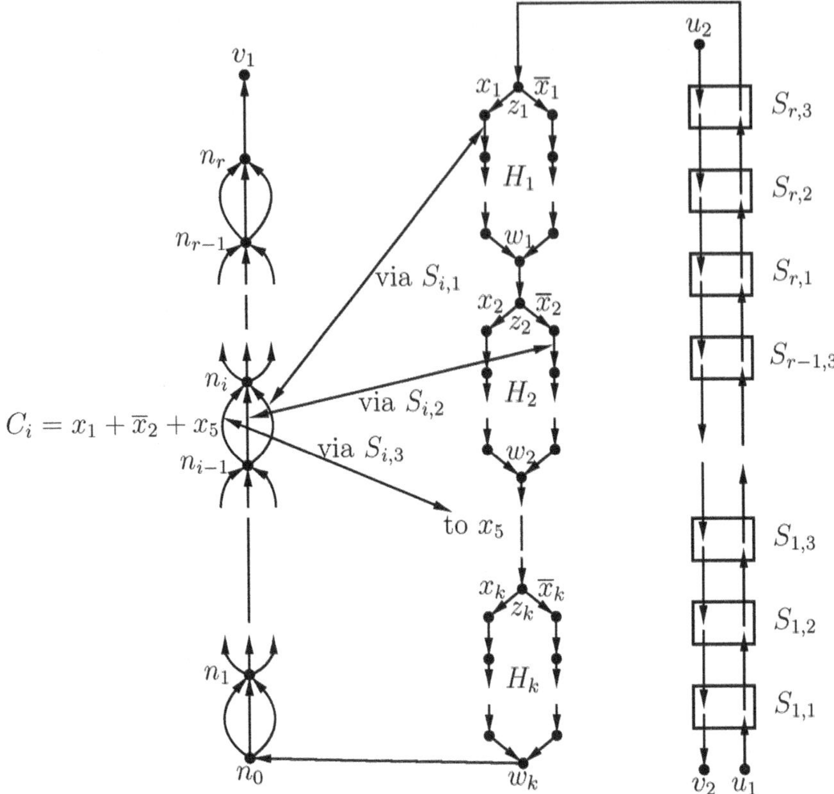

Fig. 2.4 A schematic picture of the digraph $D[\mathcal{F}]$ [5, 26]

two different arcs indicate that only one of these arcs (which are actually paths) can be used in a solution.

Finally, we join the D arc of the switch $S_{r,3}$ to the vertex z_1 of H_1, add an arc from w_k in H_k to n_0 and choose vertices u_1, u_2, v_1, v_2 as shown in the figure.

We now construct an Eulerian digraph D' from $D[\mathcal{F}]$ as follows: Firstly, for each switch $S_{i,j}$, we duplicate the arcs $A, B, C, D, 9', 9, 4', 4, 2, 2'$ (note that in this procedure there are two parallel arcs between $S_{r,3}$ and the vertex z_1). Secondly, we duplicate the arc $w_k n_0$, and the arc $w_i z_{i+1}$ for each $i \in [k-1]$, respectively; we also duplicate the arc $n_r v_1$ twice, and add the arc $v_1 n_0$. Observe that now each u_i has exactly two arcs into $U = V(D[\mathcal{F}]) \setminus \{u_1, u_2, v_1, v_2\}$ and has no arc from it, v_1 has three arcs from U and one arc into it, and v_2 has two arcs from U and has no arc into it. Finally, we add a new vertex x and the following new arcs: add the arcs $v_i x$ and $x u_j$ twice for each $1 \le i, j \le 2$. To avoid parallel arcs, we could subdivide each new arc.

It is not difficult to show that $D[\mathcal{F}]$ contains a pair of vertex-disjoint (u_1, v_1)-, (u_2, v_2)-paths if and only if D' contains a pair of vertex-disjoint (u_1, v_1)-, (u_2, v_2)-

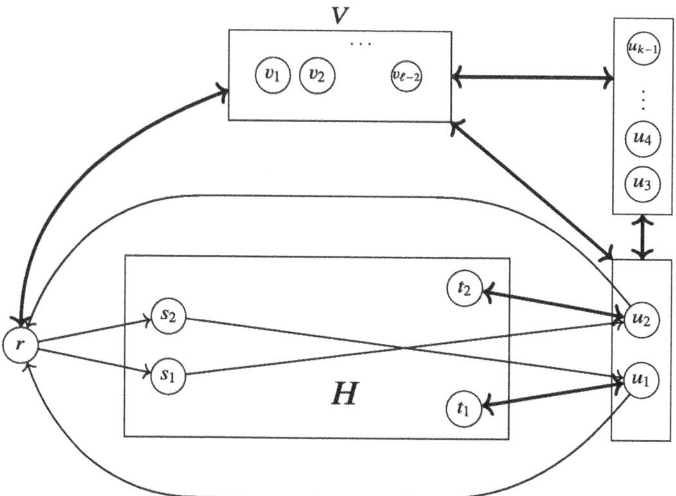

Fig. 2.5 Illustration of D^* in the proof of Theorem 2.14

paths. Indeed, if $D[\mathcal{F}]$ contains a pair of vertex-disjoint (u_1, v_1)-, (u_2, v_2)-paths, then clearly these two paths are also vertex-disjoint (u_1, v_1)-, (u_2, v_2)-paths in D'. For the other direction, let P and Q be vertex-disjoint (u_1, v_1)-, (u_2, v_2)-paths in D', then clearly they do not contain arcs incident to x and the arc v_1n_0. If they use other new arcs, then we can replace them by the corresponding parallel original arcs in $D[\mathcal{F}]$, and obtain desired paths in $D[\mathcal{F}]$. It was proved in [26] that $D[\mathcal{F}]$ contains a pair of vertex-disjoint (u_1, v_1)-, (u_2, v_2)-paths if and only if $\mathcal{F}$ is satisfiable. Hence, the Eulerian digraph D' contains a pair of vertex-disjoint (u_1, v_1)-, (u_2, v_2)-paths if and only if $\mathcal{F}$ is satisfiable, and therefore the 2-linkage problem for Eulerian digraphs is NP-complete. □

Using Theorem 2.13, we can prove the following result, which completes all the entries in Table 2.6.

Theorem 2.14 ([73]) *Let $k \geq 3$ and $\ell \geq 2$ be fixed integers. Let D be an Eulerian digraph and $S \subseteq V(D)$ with $|S| = k$ and $r \in S$. Then deciding whether $\kappa_{S,r}(D) \geq \ell$ is NP-complete.*

Proof It is not difficult to see that the problem belongs to NP. We will show that the problem is NP-hard by reducing from DIRECTED 2-LINKAGE in Eulerian digraphs. Let H be an Eulerian digraph and let s_1, s_2, t_1, t_2 be distinct vertices in H. We now produce a new Eulerian digraph D^* with $V(D^*) = V(H) \cup V \cup U \cup \{r\}$, where $V = \{v_1, v_2, \ldots v_{\ell-2}\}$ and $U = \{u_1, u_2, \ldots, u_{k-1}\}$ (here let $S = \{r, u_1, u_2, \ldots, u_{k-1}\}$). Furthermore let the arcs set of D^* be defined as follows (see Fig. 2.5).

Table 2.6 Eulerian digraphs

$\kappa_{S,r}(D) \geq \ell$? $\lvert S\rvert = k$	$k = 3$	$k \geq 4$ constant	k part of input
$\ell = 2$	NP-complete [73]	NP-complete [73]	NP-complete [73]
$\ell \geq 3$ constant	NP-complete [73]	NP-complete [73]	NP-complete [73]
ℓ part of input	NP-complete [73]	NP-complete [73]	NP-complete [73]

$$\begin{aligned} A(D^*) = A(H) &\cup \{rs_1, rs_2, t_1u_1, u_1t_1, t_2u_2, u_2t_2, s_1u_2, s_2u_1, u_1r, u_2r\} \\ &\cup \{rv, vr, vu, uv \mid v \in V \text{ and } u \in U\} \\ &\cup \{u_iu_j, u_ju_i \mid i \in [2] \text{ and } j = 3, 4, \ldots, k-1\} \end{aligned}$$

Note that D^* is Eulerian and let $S = \{r\} \cup U$. We will show that $\kappa_{S,r}(D^*) \geq \ell$ if and only if there exist two vertex-disjoint paths, P_1 and P_2, such that P_i is an (s_i, t_i)-path, for $i \in [2]$. This will complete the proof.

First assume that there exist two vertex-disjoint paths, P_1 and P_2, such that P_i is an (s_i, t_i)-path, for $i \in [2]$. Add the arcs, rs_1, s_1u_2, t_1u_1 and all arcs from u_1 to $\{u_3, u_4, \ldots, u_{k-1}\}$ to P_1 and call the resulting (S, r)-tree for $T_{\ell-1}$. Analogously, add the arcs, rs_2, s_2u_1, t_2u_2 and all arcs from u_2 to $\{u_3, u_4, \ldots, u_{k-1}\}$ to P_2 and call the resulting (S, r)-tree for T_ℓ. Finally let T_i be the (S, r)-tree containing the arc rv_i and all arcs from v_i to U, for each $i \in [\ell - 2]$. The (S, r)-trees, $T_1, T_2, \ldots, T_\ell$ are now internally-disjoint, which implies that $\kappa_{S,r}(D^*) \geq \ell$ as desired.

Conversely assume that $\kappa_{S,r}(D^*) \geq \ell$ and let T_1 and T_2 be two (S, r)-trees in D^* that do not use any of the vertices in V (which exist as $|V| = \ell - 2$). Without loss of generality assume that $rs_1 \in A(T_1)$ and $rs_2 \in A(T_2)$. As the arcs into u_1 in T_1 and T_2 is either s_2u_1 or t_1u_1, we note that $s_2u_1 \in A(T_2)$ (as $s_2 \in V(T_2)$) and $t_1u_1 \in A(T_1)$. Analogously, $s_1u_2 \in A(T_1)$ and $t_2u_2 \in A(T_2)$. This implies that there is an (s_1, t_1)-path in T_1 and an (s_2, t_2)-path in T_2, which are vertex-disjoint. This completes the proof of the fact that deciding whether $\kappa_{S,r}(D) \geq \ell$ is NP-complete. □

It may be slightly surprising that for Eulerian digraphs the complexity of deciding if $\lambda_{S,r}(D) \geq \ell$ is always polynomial, while the complexity of deciding if $\kappa_{S,r}(D) \geq \ell$ is always NP-complete.

It would also be interesting to determine the complexity for other classes of digraphs, like semicomplete digraphs. For example, one may consider the following question.

Problem 2.1 ([73]) What is the complexity of deciding whether $\kappa_{S,r}(D) \geq \ell$ (resp. $\lambda_{S,r}(D) \geq \ell$) for integers $k \geq 3$ and $\ell \geq 2$, and a semicomplete digraph D?

In the argument for the question of deciding whether $\kappa_{S,r}(D) \geq \ell$ for a symmetric digraph D when both k and ℓ are fixed, Sun and Yeo [73] used Corollary 2.1 where the $2k$ vertices $s_1, s_2, \ldots, s_k, t_1, t_2, \ldots, t_k$ are not necessarily distinct. Normally in the k-linkage problem the initial and terminal vertices are considered distinct. However if we allow them to be non-distinct and look for internally-disjoint paths instead of vertex-disjoint paths, and the problem remains polynomial, then a similar approach to that of Theorem 2.6 and Corollary 2.2 can be used to show polynomiality of deciding whether $\kappa_{S,r}(D) \geq \ell$.

2.3 Related Topics

2.3.1 Directed Tree Connectivity

The following concept of directed tree connectivity is related to directed Steiner tree packing problem and is a natural extension of tree connectivity [38, 53, 54] of undirected graphs to directed graphs. The *generalized k-vertex-strong connectivity* of D is defined as

$$\kappa_k(D) = \min\{\kappa_{S,r}(D) \mid S \subseteq V(D), |S| = k, r \in S\}.$$

Similarly, the *generalized k-arc-strong connectivity* of D is defined as

$$\lambda_k(D) = \min\{\lambda_{S,r}(D) \mid S \subseteq V(D), |S| = k, r \in S\}.$$

By definition, when $k = 2$, $\kappa_2(D) = \kappa(D)$ and $\lambda_2(D) = \lambda(D)$. Hence, these two parameters could be seen as generalizations of vertex-strong connectivity and arc-strong connectivity of a digraph. The generalized k-vertex-strong connectivity and k-arc-strong connectivity are also called *directed tree connectivity*.

In [73], Sun and Yeo proved some equalities and inequalities for directed tree connectivity. They studied the relation between the directed tree connectivity and classical connectivity of digraphs by showing that $\kappa_k(D) \leq \kappa(D)$ (when $n \geq k + \kappa(D)$) and $\lambda_k(D) = \lambda(D)$ (this equality provides a new definition for the classical arc-strong connectivity $\lambda(D)$ of digraphs, we call it "tree" version definition for $\lambda(D)$). Furthermore, the upper bound for $\kappa_k(D)$ is sharp. Let D be a strong digraph of order n. For $2 \leq k \leq n$, they proved that $1 \leq \kappa_k(D) \leq n - 1$ and $1 \leq \lambda_k(D) \leq n - 1$. All bounds are sharp, and they also characterized those digraphs D for which $\kappa_k(D)$ (respectively, $\lambda_k(D)$) attains the upper bound. In the same paper, sharp Nordhaus-Gaddum type bounds (see [1] for a survey on the topic of Nordhaus-Gaddum type relations) for $\lambda_k(D)$ were also given; moreover, extremal digraphs for the lower bounds were characterized.

A digraph $D = (V(D), A(D))$ is called *minimally generalized* (k, ℓ)*-vertex (resp. arc)-strongly connected* if $\kappa_k(D) \geq \ell$ (resp. $\lambda_k(D) \geq \ell$) but for any arc $e \in A(D)$, $\kappa_k(D-e) \leq \ell-1$ (resp. $\lambda_k(D-e) \leq \ell-1$), where $2 \leq k \leq n$, $1 \leq \ell \leq n-1$. In [66], Sun studied the minimally generalized (k, ℓ)-vertex-strongly connected digraphs and minimally generalized (k, ℓ)-arc-strongly connected digraphs.

2.3.2 *Independent Directed Steiner Trees and Independent Branchings*

Let D be a digraph with $r \in S \subseteq V(D)$. In an (S, r)-tree T, we use $T(r, s)$ to denote the unique path in T from the root r to s, where $s \in S$ is a terminal vertex. Two (S, r)-trees T_1 and T_2 are said *arc-independent* if the paths $T_1(r, s)$ and $T_2(r, s)$ are arc-disjoint for every terminal $s \in S$. Two (S, r)-trees T_1 and T_2 are said *internally independent* if the paths $T_1(r, s)$ and $T_2(r, s)$ are internally-disjoint for every terminal $s \in S$. If an (S, r)-tree T is a spanning subdigraph of D, then it is called an *r-branching* of D. Similarly, we can define the concepts of *arc-independent r-branchings* and *internally independent r-branchings*. There are some literature on these topics, such as [17, 29, 40–42, 85].

2.3.3 *Arc-Disjoint In- and Out-Branchings Rooted at the Same Vertex*

Recall that it is NP-complete to decide whether a digraph D has a pair of arc-disjoint out-branching and in-branching rooted at r, which was proved by Thomassen (see [3]). Following [10] we will call such a pair a *good pair rooted at r*. Note that a good pair forms a strong spanning subdigraph of D and thus if D has a good pair, then D is strong. The problem of the existence of a good pair was studied for tournaments and their generalizations, and characterizations (with proofs implying polynomial-time algorithms for finding such a pair) were obtained in [3] for tournaments, [8] for quasi-transitive digraphs and [10] for locally semicomplete digraphs (see Corollary 5.3). Also, Bang-Jensen and Huang [8] showed that if r is adjacent to every vertex of D (apart from itself) then D has a good pair rooted at r. Gutin and Sun [37] studied the existence of a good pair in compositions of digraphs. They obtained the following result: every strong digraph composition Q in which $n_i \geq 2$ for every $i \in [t]$, has a good pair at every vertex of Q. The condition of $n_i \geq 2$ in this result cannot be relaxed. They also characterized semicomplete compositions with a good pair, which generalizes the corresponding characterization by Bang-Jensen and Huang [8] for quasi-transitive digraphs. As a result, we can decide in polynomial time whether a given semicomplete composition has a good pair rooted at a given vertex. For more information on the topic of digraph compositions, the reader can see [11, 37, 70, 71, 75, 77] and a survey [65].

Chapter 3
Directed Steiner Path Packing Problem

Abstract In the two former sections, we introduce the complexity for IDSPP on Eulerian digraphs and symmetric digraphs, and the complexity for ADSPP on general digraphs, when both k and ℓ are fixed. In the third section, we introduce a related topic: directed path connectivity.

3.1 Results for IDSPP

By Theorem 2.13, Sun and Zhang [74] proved the NP-completeness of deciding whether $\kappa^p_{S,r}(D) \geq \ell$ for Eulerian digraphs (and therefore for general digraphs).

Theorem 3.1 ([74]) *Let the two integers $k \geq 3, \ell \geq 1$ be both fixed. Given an Eulerian digraph D and $r \in S \subseteq V(D)$ satisfying $|S| = k$, it is NP-complete to decide if $\kappa^p_{S,r}(D) \geq \ell$.*

In their argument, Sun and Zhang proved the NP-hardness of this problem by the NP-hardness of DIRECTED 2-LINKAGE problem in Eulerian digraphs (Theorem 2.13). Their construction of the reduction for the case that $k \geq 3, \ell \geq 2$ is as follows [74]:

Let $[H; s_1, s_2, t_1, t_2]$ be any instance of the latter problem such that H is Eulerian, and (s_1, t_1, s_2, t_2) is a sequence of four terminal vertices of H. Now let $V(H') = V(H) \cup S \cup \{u_1, u_2\}$ with $S = \{x_i \mid i \in [k]\}$, and let

$$\begin{aligned} A(H') = A(H) &\cup \{x_1s_1, t_1x_2, x_{k-1}s_2, t_2x_k, s_1u_1, u_1t_2, s_2u_2, u_2t_1\} \\ &\cup \{x_ix_{i+1} \mid i \in [k]\}, \end{aligned}$$

where $x_{k+1} = x_1$. Furthermore, replicate the arc x_ix_{i+1} $\ell - 1$ additional copies for each $i \in \{2, \ldots, k-2, k\}$, and replicate the two arcs x_1x_2 and $x_{k-1}x_k$ $\ell - 2$ additional copies. Finally, to avoid parallel arcs, insert a new vertex $z^j_{i,i+1}$ to each arc x_ix_{i+1}, where $j \in [\ell]$ if $i \in \{2, \ldots, k-2, k\}$, and $j \in [\ell - 1]$ otherwise. We use D to denote the resulting digraph with $r = x_1$ as shown in Fig. 3.1.

Y. Sun, *Steiner Type Packing Problems in Digraphs*, SpringerBriefs in Mathematics,
https://doi.org/10.1007/978-981-95-8743-8_3

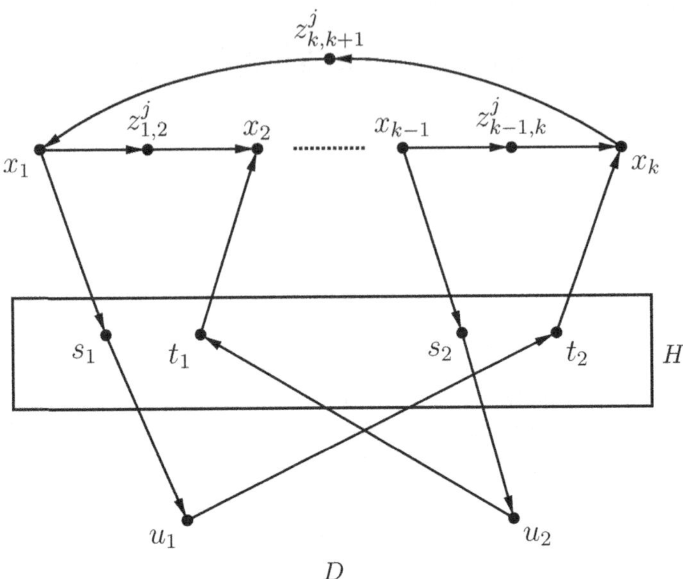

Fig. 3.1 The digraph D

However, when we consider the class of symmetric digraphs, the problem becomes polynomial-time solvable according to the following Theorem 3.2. By Corollary 2.1, we can obtain the following lemma which will be used in the argument of Theorem 3.2.

Lemma 3.1 ([74]) *Let the two integers $k \geq 3, \ell \geq 2$ be both fixed. Let D be a symmetric digraph and $r \in S \subseteq V(D)$ satisfying $|S| = k$. Let $\cup_{0 \leq i \leq \ell} A_i$ be a partition of the arc set, denoted by $A(D[S])$, in $D[S]$.*

The following problem can be solved in $O(n^3 k^{k\ell})$ time: decide whether there are ℓ pairwise internally-disjoint (S, r)-paths: P_i $(i \in [\ell])$, such that for each $i \in [\ell]$, $A(D[S]) \cap A(P_i) = A_i$.

Theorem 3.2 ([74]) *Let $k \geq 3$ and $\ell \geq 2$ be fixed integers. We can in polynomial time decide if $\kappa^p_{S,r}(D) \geq \ell$ for any symmetric digraph D with $S \subseteq V(D)$, with $|S| = k$ and $r \in S$.*

It would also be interesting to study the complexity of IDSPP on other digraph classes, such as semicomplete digraphs.

Problem 3.1 Let $k \geq 3$ and $\ell \geq 1$ be fixed integers. For a semicomplete digraph D and $r \in S \subseteq V(D)$ with $|S| = k$, what is the complexity of deciding whether $\kappa^p_{S,r}(D) \geq \ell$?

3.2 Results for ADSPP

Fortune, Hopcroft and Wyllie proved the following important theorem on DIRECTED WEAK 2-LINKAGE.

Theorem 3.3 ([26]) *The* DIRECTED WEAK 2-LINKAGE *is NP-complete.*

By Theorems 3.1 and 3.3, Sun and Zhang proved the NP-completeness of deciding whether $\lambda^p_{S,r}(D) \geq \ell$ for general digraphs.

Theorem 3.4 ([74]) *Let the two integers $k \geq 3, \ell \geq 1$ be both fixed. Given a digraph D and $r \in S \subseteq V(D)$ satisfying $|S| = k$, it is NP-complete to decide if $\lambda^p_{S,r}(D) \geq \ell$.*

In their argument, Sun and Zhang proved the NP-hardness of this problem by the NP-hardness of DIRECTED WEAK 2-LINKAGE problem (Theorem 3.3). Their construction of the reduction for the case that $k \geq 3, \ell \geq 2$ is as follows [74]:

Let $[H; s_1, s_2, t_1, t_2]$ be any instance of DIRECTED WEAK 2-LINKAGE problem such that H is a digraph, and (s_1, t_1, s_2, t_2) is a sequence of four terminal vertices of H. Let $V(H') = V(H) \cup S$ with $S = \{x_i \mid i \in [k]\}$, and let

$$A(H') = A(H) \cup \{x_1s_1, t_1x_2, x_{k-1}s_2, t_2x_k\} \cup \{x_ix_{i+1} \mid i \in [k-1]\}.$$

Furthermore, we replicate the arc x_ix_{i+1} $\ell - 1$ additional copies for each $i \in \{2, \ldots, k-2\}$, and replicate the two arcs x_1x_2 and $x_{k-1}x_k$ $\ell - 2$ additional copies. Finally, to avoid parallel arcs, insert a new vertex $z^j_{i,i+1}$ to each arc x_ix_{i+1}, where $j \in [\ell]$ if $i \in \{2, \ldots, k-2\}$, and $j \in [\ell - 1]$ otherwise. Let H'' be the resulting digraph with $r = x_1$.

It would also be interesting to study the complexity of ADSPP on some digraph classes, such as semicomplete digraphs, Eulerian digraphs and symmetric digraphs.

Problem 3.2 Let $k \geq 3$ and $\ell \geq 1$ be fixed integers. For a semicomplete digraph (Eulerian digraph, or symmetric digraph) D and $r \in S \subseteq V(D)$ with $|S| = k$, what is the complexity of deciding whether $\lambda^p_{S,r}(D) \geq \ell$?

3.3 A Related Topic: Directed Path Connectivity

The following concept of directed path connectivity is related to the directed Steiner path packing problem and is a natural extension of path connectivity of undirected graphs (see [53] for the introduction of path connectivity) to directed graphs. The *directed path k-connectivity* of D is defined as

$$\kappa^p_k(D) = \min\{\kappa^p_{S,r}(D) \mid S \subseteq V(D), |S| = k, r \in S\}.$$

Similarly, the *directed path k-arc-connectivity* of D is defined as

$$\lambda_k^P(D) = \min\{\lambda_{S,r}^P(D) \mid S \subseteq V(D), |S| = k, r \in S\}.$$

By definition, when $k = 2$, $\kappa_2^P(D) = \kappa(D)$ and $\lambda_2^P(D) = \lambda(D)$, where $\kappa(D)$ and $\lambda(D)$ are vertex-strong connectivity and arc-strong connectivity of digraphs, respectively. Hence, these two types of directed path connectivity could be seen as generalizations of classical connectivity of a digraph.

In [74], Sun and Zhang studied the parameters $\kappa_k^P(D)$ and $\lambda_k^P(D)$. They showed that the values $\lambda_k^P(D)$ is decreasing over k, but the values $\kappa_k^P(D)$ are neither increasing, nor decreasing over k. They gave sharp upper bounds for the parameters $\kappa_k^P(D)$ and $\lambda_k^P(D)$ in terms of classical connectivity $\kappa(D)$ and $\lambda(D)$ of D. Their results mean that the parameter $\lambda_k^P(D)$ (resp. $\kappa_k^P(D)$) is also a generalization of the classical edge-connectivity (resp. vertex-connectivity) of undirected graphs. Finally, they got sharp lower and upper bounds for the Nordhaus-Gaddum type relations (see [1] for a survey on the topic of Nordhaus-Gaddum type relations) of the parameter $\lambda_k^P(D)$.

Chapter 4
Strong Subgraph Packing Problem

Abstract In the following two former sections, we introduce the complexity for the decision versions of ISSP or/and ASSP on general digraphs, semicomplete digraphs, symmetric digraphs and Eulerian digraphs. Inapproximability results on ISSP and ASSP are mentioned in Sect. 4.3. The last section, Sect. 4.4, is about two related topics. The first one includes Kriesell conjecture and its extension in strong subgraph packing problem, and the second one concerns a new connectivity on digraphs, strong subgraph connectivity.

4.1 General Digraphs

For a fixed $k \geq 2$, it is easy to decide whether $\kappa_S(D) \geq 1$ for a digraph D with $|S| = k$: it holds if and only if D is strong. Unfortunately, deciding whether $\kappa_S(D) \geq 2$ is already NP-complete for $S \subseteq V(D)$ with $|S| = k$, where $k \geq 2$ is a fixed integer.

By using the reduction from the DIRECTED k-LINKAGE problem (Theorem 2.1), we can prove the following intractability result.

Theorem 4.1 ([76]) *Let $k \geq 2$ and $\ell \geq 2$ be fixed integers. Let D be a digraph and $S \subseteq V(D)$ with $|S| = k$. The problem of deciding whether $\kappa_S(D) \geq \ell$ is NP-complete.*

Proof Clearly, the problem is in NP. To show it is NP-hard, we reduce from the DIRECTED 2-LINKAGE problem, which is NP-complete (Theorem 2.1).

Let us first consider the case of $\ell = 2$ and $k = 2$. Let $[D; s_1, t_1, s_2, t_2]$ be an instance of DIRECTED 2-LINKAGE. Let us construct a new digraph D' (see Fig. 4.1) by adding to D vertices x, y and arcs

$$t_1x, xs_1, t_2y, ys_2, xs_2, s_2x, yt_1, t_1y.$$

Let $S = \{x, y\}$. It remains to show that $[D; s_1, t_1, s_2, t_2]$ is a positive instance of DIRECTED 2-LINKAGE if and only if $\kappa_S(D') \geq 2$.

Let $[D; s_1, t_1, s_2, t_2]$ be a positive instance of DIRECTED 2-LINKAGE with vertex-disjoint paths P_1, P_2 from s_1 to t_1 and from s_2 to t_2, respectively. Then there

Y. Sun, *Steiner Type Packing Problems in Digraphs*, SpringerBriefs in Mathematics,
https://doi.org/10.1007/978-981-95-8743-8_4

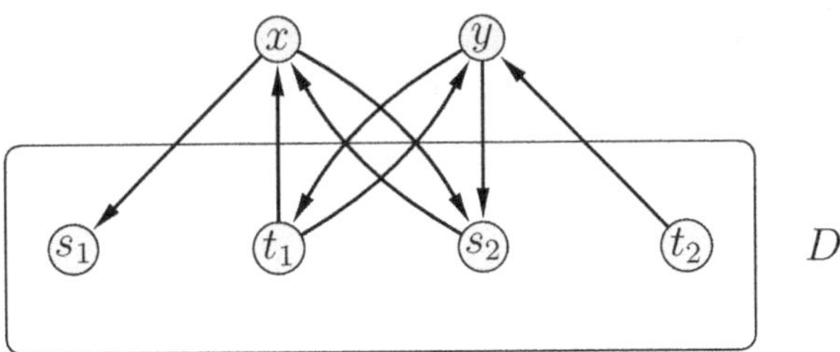

Fig. 4.1 The digraph D'

are two internally-disjoint S-strong subgraphs of D', one induced by the arcs of P_1 and t_1x, xs_1, t_1y, yt_1 and the other by the arcs of P_2 and t_2y, ys_2, xs_2, s_2x.

Let D' have two internally-disjoint S-strong subgraphs H_1, H_2. Since the in-degree of x in D' is 2, we may without loss of generality assume that $t_1 \in V(H_1)$ and $s_2 \in V(H_2)$. As y has in-degree 2 and $t_1 \in V(H_1)$ we must have $t_2 \in V(H_2)$. As the out-degree of x is 2, we analogously have $s_1 \in V(H_1)$ (as $s_2 \in V(H_2)$). So, for $i \in [2]$, both s_i and t_i are in H_i. Therefore, there must be a path P_i from s_i to t_i in H_i and by definition of D', P_i will not have vertices outside of D. As H_1 and H_2 are internally-disjoint, the paths are disjoint.

Now let us consider the case of $\ell \geq 3$ and $k = 2$. Add to D' $\ell - 2$ copies of the 2-cycle xyx and subdivide the arcs of every copy to avoid parallel arcs. Let us denote the new digraph by D''. Assume that there are ℓ internally-disjoint S-strong subgraphs, $H_1, H_2, \ldots H_\ell$, in D''. As the out-degree of y in D'' is ℓ we can without loss of generality assume that $t_1 \in V(H_1)$, $s_2 \in V(H_2)$ and the $\ell - 2$ (subdivided) arcs from y to x belong to $H_3, H_4, \ldots, H_\ell$, respectively. As $t_1 \in V(H_1)$ and the in-degree of y is ℓ no (subdivided) arc from x to y belongs to H_1. Analogously, since $s_2 \in V(H_2)$ and the out-degree of x is ℓ, no (subdivided) arc from x to y belongs to H_2. Therefore the (subdivided) arcs from x to y belong to $H_3, H_4, \ldots, H_\ell$, respectively. As in the case when $\ell = 2$ we now note that $s_1 \in V(H_1)$ and $t_2 \in V(H_2)$ and that there therefore exist two disjoint paths from s_1 to t_1 and s_2 to t_2 in D, respectively.

Conversely if there exist two disjoint paths from s_1 to t_1 and s_2 to t_2 in D, then it is not difficult to create ℓ internally-disjoint S-strong subgraphs in D'' using the same approach as when $\ell = 2$ as each (subdivided) 2-cycle xyx also gives rise to an S-strong subgraph. Thus, we have proved the theorem in the case of $k = 2$ and $\ell \geq 2$.

It remains to consider the case of $\ell \geq 2$ and $k \geq 3$. Add to D'' (where $D'' = D'$ for $\ell = 2$) $k - 2$ new vertices $x_1, \ldots, x_{k-2}$ and arcs of ℓ 2-cycles xx_ix for each $i \in [k-2]$. Subdivide the new arcs to avoid parallel arcs. Let $S = \{x, y, x_1, \ldots, x_{k-2}\}$. It is not hard to see that the resulting digraph has ℓ internally-disjoint S-strong subgraphs if and only if $[D; s_1, t_1, s_2, t_2]$ is a positive instance of Directed 2-Linkage. □

Yeo proved that it is an NP-complete problem to decide whether a 2-regular digraph has two arc-disjoint Hamiltonian cycles (see, e.g., Theorem 6.6 in [12]).

Thus, the problem of deciding whether $\lambda_{V(D)}(D) \geq 2$ is NP-complete. Sun and Gutin [68] extended this result in Theorem 4.2 which can be proved by Theorem 3.3.

Theorem 4.2 ([68]) *Let $k \geq 2$ and $\ell \geq 2$ be fixed integers. Let D be a digraph and $S \subseteq V(D)$ with $|S| = k$. The problem of deciding whether $\lambda_S(D) \geq \ell$ is NP-complete.*

Proof Clearly, the problem is in NP. We will show that it is NP-hard using a reduction similar to that in Theorem 4.1. Let us first deal with the case of $\ell = 2$ and $k = 2$. Consider the digraph D' used in the proof of in Theorem 4.1 (see Fig. 4.1), where D is an arbitrary digraph, x, y are vertices not in D, and $t_1x, xs_1, t_2y, ys_2, xs_2, s_2x, yt_1, t_1y$ are additional arcs. To construct a new digraph D'' from D', replace every vertex u of D by two vertices u^- and u^+ such that u^-u^+ is an arc in D'' and for every $uv \in A(D)$ add an arc u^+v^- to D''. Also, for $z \in \{x, y\}$, for every arc zu in D' add an arc zu^- to D'' and for every arc uz add an arc u^+z to D''.

Let $S = \{x, y\}$. It was proved in Theorem 4.1 that $\kappa_S(D') \geq 2$ if and only if there are vertex-disjoint paths from s_1 to t_1 and from s_2 to t_2. It follows from this result and definition of D'' that $\lambda_S(D'') \geq 2$ if and only if there are arc-disjoint paths from s_1^- to t_1^+ and from s_2^- to t_2^+. Since the WEAK 2-LINKAGE PROBLEM is NP-complete, we conclude that the problem of deciding whether $\lambda_S(D'') \geq 2$ is NP-hard.

Now let us consider the case of $\ell \geq 3$ and $k = 2$. Add to D'' $\ell - 2$ copies of the 2-cycle xyx and subdivide the arcs of every copy to avoid parallel arcs. Let us denote the new digraph by D'''. Similarly to that in Theorem 4.1, we can show that $\lambda_S(D''') \geq \ell$ if and only if $\lambda_S(D'') \geq 2$.

It remains to consider the case of $\ell \geq 2$ and $k \geq 3$. Add to D''' (where $D''' = D''$ for $\ell = 2$) $k - 2$ new vertices $x_1, \ldots, x_{k-2}$ and arcs of ℓ 2-cycles xx_ix for each $i \in [k-2]$. Subdivide the new arcs to avoid parallel arcs. Denote the obtained digraph by D''''. Let $S = \{x, y, x_1, \ldots, x_{k-2}\}$. Similarly to that in Theorem 4.1, we can show that $\lambda_S(D'''') \geq \ell$ if and only if $\lambda_S(D'') \geq 2$. □

Now Theorems 4.1 and 4.2 imply the entries in Tables 4.1 and 4.2.

Table 4.1 Directed graphs

$\lambda_S(D) \geq \ell$? $\|S\| = k$	$k \geq 2$ constant	k part of input
$\ell \geq 2$ constant	NP-complete [76]	NP-complete [76]
ℓ part of input	NP-complete [76]	NP-complete [76]

Table 4.2 Directed graphs

$\kappa_S(D) \geq \ell$? $\|S\| = k$	$k \geq 2$ constant	k part of input
$\ell \geq 2$ constant	NP-complete [68]	NP-complete [68]
ℓ part of input	NP-complete [68]	NP-complete [68]

4.2 Semicomplete Digraphs, Symmetric Digraphs and Eulerian Digraphs

Chudnovsky et al. [21] proved the following powerful result on DIRECTED k-LINKAGE.

Theorem 4.3 ([21]) *Let k and c be fixed positive integers. Then the* DIRECTED k-LINKAGE *problem on a digraph D whose vertex set can be partitioned into c sets each inducing a semicomplete digraph and a terminal sequence $((s_1, t_1), \ldots, (s_k, t_k))$ of distinct vertices of D, can be solved in polynomial time.*

The following nontrivial lemma can be deduced from Theorem 4.3.

Lemma 4.1 ([76]) *Let k and ℓ be fixed positive integers. Let D be a digraph with order n and let $X_1, X_2, \ldots, X_\ell$ be ℓ vertex disjoint subsets of $V(D)$, such that $|X_i| \le k$ for all $i \in [\ell]$. Let $X = \bigcup_{i=1}^{\ell} X_i$ and assume that every vertex in $V(D)\setminus X$ is adjacent to every other vertex in D. Then we can in polynomial time decide if there exists vertex disjoint subsets $Z_1, Z_2, \ldots, Z_\ell$ of $V(D)$, such that $X_i \subseteq Z_i$ and $D[Z_i]$ is strongly connected for each $i \in [\ell]$.*

Proof Let $C_i^1, C_i^2, \ldots, C_i^{r_i}$ be the strongly connected components in $D[X_i]$, such that there is no arc from C_i^b to C_i^a for $1 \le a < b \le r_i$. We consider the following two cases.

Case 1 $D[X_i]$ has a unique initial and a unique terminal component (which can be the same component) for all $i \in [\ell]$.

Let $\mathcal{T} = \emptyset$. For each $i \in [\ell]$, do the following. If $D[X_i]$ is strongly connected then set $Z_i = X_i$ and delete X_i from D. Otherwise, contract every strong component C_i^j to a vertex c_i^j and look at all possible permutations of all subsets of $\{c_i^1, c_i^2, \ldots, c_i^{r_i}\}$ containing c_i^1 and $c_i^{r_i}$ which start with $c_i^{r_i}$ and end with c_i^1. Let $Z = (z_1, z_2, \ldots, z_r)$ be such a permutation, where $z_1 = c_i^{r_i}$, $z_r = c_i^1$ and $2 \le r \le r_i$. Now duplicate every vertex z_a to z_a^s and z_a^t, for all $a = 2, 3, \ldots, r-1$ and remove every c_i^j that does not appear in the permutation. We now add the sequence $\mathcal{T}_i = ((c_i^{r_i}, z_2^t), (z_2^s, z_3^t), (z_3^s, z_4^t), \ldots, (z_{r-1}^s, c_i^1))$ to our terminal sequence $\mathcal{T}$.

We can use Theorem 4.3 for D in order to determine if there are vertex disjoint paths satisfying our terminal sequence $\mathcal{T}$ (that is, for every $(s, t) \in \mathcal{T}$ there is a path from s to t). Indeed, Theorem 4.3 can be used for D and $\mathcal{T}$ since (i) k and ℓ are constants; (ii) $D - X$ is a semicomplete digraph, every vertex in X can be viewed a semicomplete digraph, and $|X| \le k\ell$; (iii) $|\mathcal{T}| \le 2k\ell$. If such a linkage exists (for the terminal sequence $\mathcal{T}$ of some permutations above) then let Z_i include all internal vertices on paths between the pairs of vertices in $\mathcal{T}_i$ as well as X_i itself. Now observe that $D[Z_i]$ is strongly connected and all $Z_1, Z_2, \ldots, Z_\ell$ are vertex disjoint, as desired.

We will now show that if there exists Z_i, such that $D[Z_i]$ is strongly connected and all Z_i's are vertex disjoint, then there exists a desired linkage. So, assume that such Z_i exists. As $D[Z_i]$ is strong, we note that it remains strong after contracting all strong components of $D[X_i]$ to vertices. Therefore there exists a shortest path P from the terminal strong component of $D[X_i]$ to the initial strong component of $D[X_i]$. Let the vertices on P which correspond to (contracted) strong components of $D[X_i]$ be $(z_1, z_2, \ldots, z_r)$ (in the order they appear on P) and using this as the permutation in our algorithm for the subpaths of P gives us the desired linkage between the z_i's. Doing the above for all $i \in [\ell]$ we see that our algorithm will indeed find the desired linkage (when considering the permutations constructed above).

As k and ℓ are constants, we note that there are at most a constant number of permutations to consider, so the algorithm runs in polynomial time. This completes Case 1.

Case 2 Case 1 does not hold.

We will in this case transform the problem, such that we can solve it using Case 1. For all $i \in [\ell]$ proceed as follows. Initiate a set Q as an empty set. If there is a unique initial strong component in $D[X_i]$ and a unique strong terminal component in $D[X_i]$ then let $X_i' = X_i$. If this is not the case, then let $I = \{I_1, I_2, \ldots, I_p\}$ denote the set of initial strong components in $D[X_i]$ and let $T = \{T_1, T_2, \ldots, T_q\}$ denote the set of terminal strong components in $D[X_i]$. For every $I_a \in I$ choose a vertex, $v_a \in V(D) \setminus (X \cup Q)$ such that v_a has at least one arc into the component I_a. We allow repetition of vertices in the sequence $v_1, v_2, \ldots, v_p$. (Such vertices v_j must exist if there is a set Z_i containing X_i such that $D[Z_i]$ is strong.) Analogously, for each $T_b \in T$ choose a vertex, $w_b \in V(D) \setminus (X \cup Q)$ such that w_b has at least one arc into it from the component T_b. Again we allow $w_1, w_2, \ldots, w_q$ to be not necessarily distinct. Now add vertices of $v_1, v_2, \ldots, v_p$ and $w_1, w_2, \ldots, w_q$ to Q.

If for some i we cannot choose $v_1, v_2, \ldots, v_p$ as above, we stop and consider other choices for the previous values of i. Analogously, for $w_1, w_2, \ldots, w_q$.

If we have succeeded in choosing $v_1, v_2, \ldots, v_p$ and $w_1, w_2, \ldots, w_q$ for every $i \in [\ell]$, then for each $i \in [\ell]$ we add the corresponding vertices $v_1, v_2, \ldots, v_p$ and $w_1, w_2, \ldots, w_q$ to X_i and call the resulting set X_i'. Note that $|X_i'| \leq |X_i| + (p+q) \leq 3k$.

If C is a terminal component in $D[X_i']$ then C must contain a vertex not in X_i, as otherwise C would be a terminal component of $D[X_i]$ a contradiction to X_i' containing a vertex (not in X_i and therefore not in C) that has an arc into it from C. However, as all vertices not in X_i are adjacent, this implies that there is a unique terminal strong component in $D[X_i']$. Analogously, there is a unique initial strong component in $D[X_i']$.

We now use the approach in Case 1, for all possible choices of vertices v_j and w_j for all $i \in [\ell]$. As there are at most n^k possible choices of vertices v_j and w_j for each i observe that we have to use the approach in Case 1 at most $n^{k\ell}$ times, which is a polynomial as k and ℓ are constants.

If the above algorithm finds the ℓ sets, $Z_1, Z_2, \ldots, Z_\ell$, then clearly they exist. Conversely, if the sets do exist then when $D[X_i]$ is not strong, observe that each initial strong component in $D[X_i]$ must have an arc into it from a vertex in Z_i and each terminal strong component in $D[X_i]$ must have an arc out of it to a vertex in Z_i. Picking these vertices as our vertices v_j and w_j, observe that our algorithm will indeed find sets $Z_1, Z_2, \ldots, Z_\ell$, as desired. □

Using Lemma 4.1, Sun, Gutin, Yeo and Zhang proved the following result for semicomplete digraphs.

Theorem 4.4 ([76]) *Let $k \geq 2$ and $\ell \geq 2$ be fixed integers. Let D be a semicomplete digraph and $S \subseteq V(D)$ with $|S| = k$. The problem of deciding whether $\kappa_S(D) \geq \ell$ is polynomial-time solvable.*

Proof Let $k, \ell \geq 2$ be fixed and let $S = \{s_1, \ldots, s_k\}$ be a set of vertices of a semicomplete digraph D.

Let $A_1, A_2, \ldots, A_\ell$ be a partition of the arcs in $D[S]$, where some sets may be empty. That is, every arc in $D[S]$ belongs to exactly one A_i. Let D^* be obtained from D by replacing every s_i by ℓ copies, i.e. replacing s_i with $S_i = \{x_i^1, x_i^2, \ldots, x_i^\ell\}$ for all $i \in [\ell]$. Let $X_i = \{x_1^i, x_2^i, \ldots, x_k^i\}$ for all $i \in [\ell]$. If $s_i y$ is an arc from S to $V(D) \setminus S$, then $x_i^a y$ is in D^* for all $a \in [\ell]$. Analogously, if ys_i is an arc from $V(D) \setminus S$ to S, then yx_i^a is in D^* for all $a \in [\ell]$. For each $i \in [\ell]$ add the arcs of A_i to $D^*[X_i]$. That is, if $s_a s_b \in A_i$ then add the arc $x_a^i x_b^i$ to D^*. This completes the construction of D^* (for a given partition $A_1, A_2, \ldots, A_\ell$).

We can now decide if there exist disjoint vertex sets Z_i in D^* such that $X_i \subseteq Z_i$ and $D^*[Z_i]$ is strongly connected for all $i \in [\ell]$ in polynomial time by Lemma 4.1. If, for some partition, $A_1, A_2, \ldots, A_\ell$, such Z_i's exist then we will show that $\kappa_S(D) \geq \ell$ and if this is not the case then we will show that $\kappa_S(D) < \ell$. As there are only a polynomial number of partitions $A_1, A_2, \ldots, A_\ell$ (as ℓ and k are constants), this gives us a polynomial algorithm.

First assume that such Z_i's exist for some partition, $A_1, A_2, \ldots, A_\ell$. Then the subgraph in D on vertex set Z_i and with the arcs $(A(D[Z_i]) \setminus A(D[S])) \cup A_i$ is strongly connected and as all Z_i's are vertex disjoint (and the arc sets A_i's are disjoint) observe that $\kappa_S(D) \geq \ell$, as desired.

Conversely if $\kappa_S(D) \geq \ell$, then there exists strongly connected subgraphs $Y_1, Y_2, \ldots, Y_\ell$ such that $V(Y_i) \cap V(Y_j) = S$ for all $i \neq j$. Without loss of generality, we may assume that every arc of $D[S]$ belongs to some Y_i (as otherwise just add it to some Y_i). Letting $A_i = Y_i[S]$ and $Z_i = V(Y_i)$ observe that our algorithm does find the desired Z_i's and we are done. □

For semicomplete digraphs, we have the following entries of Table 4.3.

Problem 4.1 Complete the blanks in Table 4.3.

Table 4.3 Semicomplete digraphs

$\kappa_S(D) \geq \ell$? $\lvert S\rvert = k$	$k = 2$	$k \geq 3$ constant	k part of input
$\ell \geq 2$ constant	Polynomial [76]	Polynomial [76]	
ℓ part of input			

The DIRECTED k-LINKAGE problem is polynomial-time solvable for planar digraphs [61] and digraphs of bounded directed treewidth [46]. However, it seems that we cannot use the approach in proving Theorem 4.4 directly as the structure of minimum-size strong subgraphs in these two classes of digraphs is more complicated than in semicomplete digraphs. Certainly, we cannot exclude the possibility that decide if $\kappa_S(D) \geq \ell$ in planar digraphs and/or in digraphs of bounded directed treewidth is NP-complete.

Problem 4.2 What is the complexity of deciding whether $\kappa_S(D) \geq \ell$ for (fixed) integers $k \geq 2$ and $\ell \geq 2$, and a planar digraph D, where $S \subseteq V(D)$ with $|S| = k$?

Problem 4.3 What is the complexity of deciding whether $\kappa_S(D) \geq \ell$ for (fixed) integers $k \geq 2$ and $\ell \geq 2$, and a digraph D of bounded directed treewidth, where $S \subseteq V(D)$ with $|S| = k$?

It would be interesting to identify large classes of digraphs for which the $\kappa_S(D) \geq \ell$ problem can be decided in polynomial time.

Now restricted to symmetric digraphs D, for any fixed integer $k \geq 3$, by Lemma 2.1, the problem of deciding whether $\kappa_S(D) \geq \ell$ $(\ell \geq 1)$ is NP-complete for $S \subseteq V(D)$ with $|S| = k$.

Theorem 4.5 ([76]) *For any fixed integer $k \geq 3$, given a symmetric digraph D, a k-subset S of $V(D)$ and an integer ℓ $(\ell \geq 1)$, deciding whether $\kappa_S(D) \geq \ell$, is NP-complete.*

Proof It is easy to see that this problem is in NP. We divide our proof into two steps:

In the first step, let G be a tripartite graph with 3-partition $(\overline{U}, \overline{V}, \overline{W})$ such that $|\overline{U}| = |\overline{V}| = |\overline{W}| = \ell$. We will construct a graph H, a k-subset $S \subseteq V(H)$ and an integer ℓ such that there are ℓ internally-disjoint S-trees in H if and only if G is a positive instance of the CLLM PROBLEM.

We define H as follows: let $V(H) = V(G) \cup \{x_j \mid j \in [k]\}$ and $E(H) = E(G) \cup \{x_j u \mid j \in [k-2], u \in \overline{U}\} \cup \{x_{k-1} v \mid v \in \overline{V}\} \cup \{x_k w \mid w \in \overline{W}\}$. Set $S = \{x_j \mid j \in [k]\}$.

If there are ℓ internally-disjoint S-trees in H, then each tree contains exactly a vertex from $\overline{U}$, a vertex from $\overline{V}$ and a vertex from $\overline{W}$ since $\deg_H(x_i) = \ell$ for all $i \in [k]$. Furthermore, in each such tree, elements of $\{x_i \mid i \in [k-2]\}$ have exactly one common neighbor in $\overline{U}$. Since these ℓ trees are internally-disjoint, there is a partition of $V(G)$ into ℓ disjoint sets $V_1, V_2, \dots, V_\ell$ each having three vertices, such that for every $V_i = \{v_{i_1}, v_{i_2}, v_{i_3}\}$ we have that $v_{i_1} \in \overline{U}$, $v_{i_2} \in \overline{V}$, $v_{i_3} \in \overline{W}$, and $G[V_i]$ is connected.

If there is a partition of $V(G)$ into ℓ disjoint sets $V_1, V_2, \dots, V_\ell$ each having three vertices, such that for every $V_i = \{v_{i_1}, v_{i_2}, v_{i_3}\}$ we have $v_{i_1} \in \overline{U}$, $v_{i_2} \in \overline{V}$, $v_{i_3} \in \overline{W}$, and $G[V_i]$ is connected, then let T_i be a spanning tree of $G[V_i]$ together with the edge set $\{x_j v_{i_1} \mid j \in [k-2]\} \cup \{x_{k-1} v_{i_2}\} \cup \{x_k v_{i_3}\}$. It is easy to see that $T_1, T_2, \dots, T_\ell$ are the desired internally-disjoint S-trees.

In the second step, we construct a symmetric digraph D from H by replacing each edge with the corresponding arcs of both directions. If there are ℓ internally-

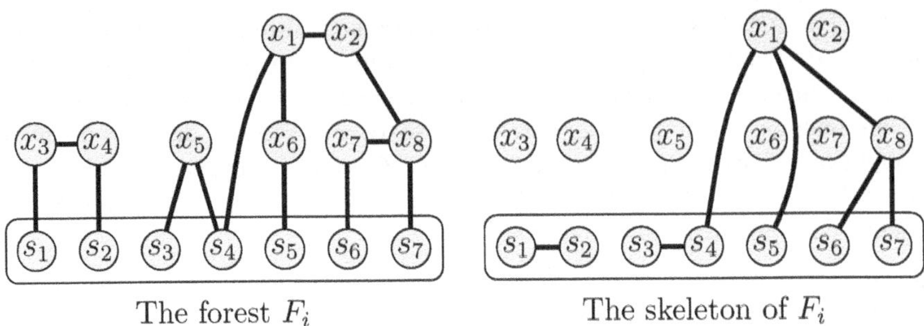

Fig. 4.2 An example of the skeleton of a forest F_i, where $S = \{s_1, s_2, s_3, s_4, s_5, s_6, s_7\}$

disjoint S-trees in H, then for each such tree, we can get an S-strong subgraph in D by replacing each edge with the corresponding arcs of both directions. Clearly, all these S-strong subgraphs of D are internally-disjoint. If there are ℓ internally-disjoint S-strong subgraphs, say D_i $(i \in [\ell])$, in D, then each D_i contains exactly a vertex from $\overline{U}$, a vertex from $\overline{V}$ and a vertex from $\overline{W}$ since $|\overline{U}| = |\overline{V}| = |\overline{W}| = \ell$. For every $i \in [\ell]$, let T_i be a spanning tree of the underlying undirected graph of D_i. Observe that $T_1, \ldots, T_\ell$ are internally-disjoint S-trees in H. We now have that there are ℓ internally-disjoint S-trees in H if and only if there are ℓ internally-disjoint S-strong subgraphs in D.

Now, by Lemma 2.1 and the two steps above, we are done. □

The last theorem assumes that k is fixed but ℓ is a part of input. When both k and ℓ are fixed, the problem of deciding whether $\kappa_S(D) \geq \ell$ for a symmetric digraph D, is polynomial-time solvable. We will start with the following technical lemma.

Lemma 4.2 ([76]) *Let $k, \ell \geq 2$ be fixed. Let G be a graph and let $S \subseteq V(G)$ be an independent set in G with $|S| = k$. For $i \in [\ell]$, let D_i be any set of arcs with both end-vertices in S. Let a forest F_i in G be called* (S, D_i)-acceptable *if the digraph $\overleftrightarrow{F_i} + D_i$ is strong and contains S. In polynomial time, we can decide whether there exists edge-disjoint forests $F_1, F_2, \ldots, F_\ell$ such that F_i is (S, D_i)-acceptable for all $i \in [\ell]$ and $V(F_i) \cap V(F_j) \subseteq S$ for all $1 \leq i < j \leq \ell$.*

Proof Assume that there exists a set $\mathcal{F} = \{F_1, F_2, \ldots, F_\ell\}$ of required forests. Observe that if there is a leaf $v \notin S$ in a forest F_i, v can be deleted from F_i and $\mathcal{F}$ will remain the required set. Thus, we may assume that all leaves in $\mathcal{F}$ are vertices of S.

Below we will use the fact that in a tree without degree-2 vertices, the number of internal vertices is smaller than the number of leaves. This fact can be easily proved by induction by deleting a leaf. Let T be a tree in a forest of $\mathcal{F}$ and let T' be the tree obtained from T by suppressing all degree-2 vertices not belonging to S. We will call T' the *skeleton* of T. Note that T' may contain edges, that are not edges of G (see Fig. 4.2 for an example). By producing the skeleton of every tree of F_i, we obtain the *skeleton* of the forest F_i.

We will now bound the number of possible skeletons obtained from F_i's. By Cayley's formula, the number of distinct trees on n labeled vertices is bounded by n^{n-2}. By considering every tree on $n_T \leq 2|S| - 1 = 2k - 1$ vertices, then assigning the n_T vertices to vertices of G and finally deleting a subset of edges in the tree, we obtain a forest in G. Note that after deleting isolated vertices every skeleton obtained from an F_i is created this way. Therefore the number of possible skeletons is bounded by

$$n_F = (n_T)^{n_T-2} \times |V(G)|^{n_T} \times 2^{n_T-1}.$$

Note that the above number is a polynomial in $|V(G)|$ as k is considered constant, which implies that n_T is constant. Thus, the number of distinct skeletons for the set $\mathcal{F}$ is bounded by n_F^{ℓ}, which is still a polynomial in $|V(G)|$ as ℓ is also considered to be a constant.

We now consider a set of skeletons $\mathcal{F}' = \{F_1', F_2', \ldots, F_\ell'\}$ where F_i' is (S, D_i)-acceptable and $V(F_i') \cap V(F_j') = S$ for all $1 \leq i < j \leq \ell$. It remains to show that there is a polynomial-time algorithm for deciding whether there exists a set $Q = \{Q_1, Q_2, \ldots, Q_\ell\}$ of required forests such that F_i' is the skeleton of Q_i. To obtain such an algorithm, we will use the celebrated result of Robertson and Seymour [60] that the UNDIRECTED p-LINKAGE problem is polynomial-time solvable.

For every forest $F_i' \in \mathcal{F}'$ and every $x \in V(F_i')$ make $d_{F_i'}(x)$ copies of x in G, such that all copies have the same neighbourhood as x. If x belongs to several forests in $\mathcal{F}'$ (which can happen if $x \in S$) then do the above for every forest, implying that we increase the number of copies of x several times. Let U_x denote all copies of x. Note that if $xy \in E(G)$ then all vertices in U_x are adjacent to all vertices in U_y. Finally, for each edge uv, add a new vertex z_{uv} and replace every edge ab between U_x and U_y by path $az_{ab}b$. Let us denote the resulting graph by G^*.

It remains to solve the instance of the (polynomial-time solvable) UNDIRECTED p-LINKAGE problem given by G^* and the terminal sequence $\{(x, y) \mid xy \in E(F_1) \cup \cdots \cup E(F_\ell)\}$. If the instance is a Yes-instance, then the found vertex-disjoint paths correspond to vertex-disjoint paths in G (as guaranteed by vertices z_{xy}) and vice versa. Therefore, if all of these vertex-disjoint paths exist, then they give us the desired set Q and if they do not exist then the desired set Q does not exist. □

Sun, Gutin, Yeo and Zhang proved the following result by Lemma 4.2:

Theorem 4.6 ([76]) *Let $k, \ell \geq 2$ be fixed. For any symmetric digraph D and $S \subseteq V(D)$ with $|S| = k$ we can in polynomial time decide whether $\kappa_S(D) \geq \ell$.*

Proof Let k, ℓ, D and S be defined as in the statement of the theorem. Let A_S be the set of arcs in $D[S]$. As $|S| = k$ we note that $|A_S| \leq 2\binom{k}{2}$.

Let $\mathcal{P} = \{P_1, P_2, \ldots, P_\ell\}$ be any partition of A_S (i.e., all sets of $\mathcal{P}$ are disjoint and their union is A_S; some sets of $\mathcal{P}$ may be empty). Let G_S be the underlying undirected graph of $D - A_S$. We can now use Lemma 4.2 to determine if there exist edge-disjoint forests $F_1, F_2, \ldots, F_\ell$ in G_S such that F_i is (S, P_i)-acceptable for all $i \in [\ell]$ and $V(F_i) \cap V(F_j) \subseteq S$ for all $1 \leq i < j \leq \ell$.

If such a set of forests exist then we will show that $\kappa_S(D) \geq \ell$ and if such a set of forests do not exist for any partition $\mathcal{P}$ then we will show that $\kappa_S(D) < \ell$. Lemma 4.2 and the fact that the number of partitions is bounded by $|A_S|^\ell$ and $|A_S| \leq 2\binom{k}{2}$ would imply the desired polynomial algorithm (as k and ℓ are fixed).

First assume that the set of forests, $F_1, F_2, \ldots, F_\ell$, exist. Let $H_i = \overleftrightarrow{F_i} + P_i$. By definition of (S, P_i)-acceptability we observe that H_i is strongly connected and $S \subseteq V(H_i)$. Furthermore we observe that $V(H_i) \cap V(H_j) = S$ and $A(H_i) \cap A(H_j) = \emptyset$ for all $1 \leq i < j \leq \ell$ and therefore $\kappa_S(D) \geq \ell$.

We will now show that if $\kappa_S(D) \geq \ell$ then the forests $F_1, F_2, \ldots, F_\ell$ do exist for some partition $\mathcal{P}$. This will complete the proof. Assume that $\kappa_S(D) \geq \ell$ and let $H_1, H_2, \ldots, H_\ell$ be strong subgraphs in D such that $V(H_i) \cap V(H_j) = S$ and $A(H_i) \cap A(H_j) = \emptyset$ for all $1 \leq i < j \leq \ell$. Let P_i^* be the arcs from A_S that belong to H_i. Let $H_i' = H_i - P_i^*$ and let L_i be the undirected underlying graph of H_i'. In each connected component of L_i choose a spanning tree. It remains to observe that union of the complete biorientations of the trees plus P_i^* is strong since H_i is strong and each spanning tree "preserves" connectivity of its component of L_i. □

Recall that it was proved in [76] that $\kappa_2(\overleftrightarrow{G}) = \kappa(G)$, which means that $\kappa_2(\overleftrightarrow{G})$ can be computed in polynomial time. In fact, the argument also means that $\kappa_{\{x,y\}}(\overleftrightarrow{G}) = \kappa_{\{x,y\}}(G)$, that is, the maximum number of disjoint (x, y)-paths in G, therefore can be computed in polynomial time. Then combining with Theorems 4.5–4.7 (which is deduced from Theorem 2.8 by constructing a reduction from the problem of 2-COLORING HYPERGRAPHS), we can complete all the entries of Table 4.4.

Theorem 4.7 ([77]) *For any fixed integer $\ell \geq 2$, given a symmetric digraph D, a k-subset S of $V(D)$ and an integer k ($k \geq 2$), deciding whether $\kappa_S(D) \geq \ell$, is NP-complete.*

Proof It is easy to see that this problem is in NP. We will reduce from the NP-complete problem of 2-COLORING HYPERGRAPHS (see [55]). That is, we are given a hypergraph H with vertex set $V(H)$ and edge set $E(H)$, and want to determine if we can 2-colour the vertices $V(H)$ such that every hyperedge in $E(H)$ contains vertices of both colours. This problem can also be formulated in terms of a bipartite graph G that has sides X (vertices) and Y (sets of vertices) such that for $x \in X$, $y \in Y$, $xy \in E(G)$ if and only if $x \in y$, and we want to 2-color X such that each vertex in Y has neighbors of both colors.

Let $G = (X \cup Y, E(G))$ be an instance of the problem of 2-COLORING HYPERGRAPHS. Define a symmetric digraph D as follows. Let $U = \{u_1, u_2, \ldots, u_{\ell-2}\}$ and let $V(D) = X \cup Y \cup U \cup \{r\}$ and let the arc set of D be defined as follows.

$$\begin{aligned} A(D) = &\{xy, yx \mid x \in X,\ y \in Y \text{ and } x \in y\} \\ &\cup\ \{ru_i, u_ir, u_iy, yu_i \mid u_i \in U \text{ and } y \in Y\} \\ &\cup\ \{rx, xr \mid x \in X\} \end{aligned}$$

Table 4.4 Symmetric digraphs

$\kappa_S(D) \geq \ell$? $\lvert S\rvert = k$	$k = 2$	$k \geq 3$ constant	k part of input
$\ell \geq 2$ constant	Polynomial [76]	Polynomial [76]	NP-complete [77]
ℓ part of input	Polynomial [76]	NP-complete [76]	NP-complete [76]

Table 4.5 Eulerian digraphs

$\kappa_S(D) \geq \ell$? $\lvert S\rvert = k$	$k = 2$	$k \geq 3$ constant	k part of input
$\ell \geq 2$ constant	NP-complete [77]	NP-complete [77]	NP-complete [77]
ℓ part of input	NP-complete [77]	NP-complete [77]	NP-complete [77]

Let $S = Y \cup \{r\}$. This completes the construction of D and S. We will show that $\kappa_S(D) \geq \ell$ if and only if G is 2-colourable, which will complete the proof.

First assume that G is 2-colourable and let R be the red vertices in G and B be the blue vertices in G in a proper 2-colouring of G. For $i \in [\ell - 2]$, let D_i contain all arcs between u_i and S. Let $D_{\ell-1}$ contain all arcs between r and R, and for each $y \in Y$ we add all arcs between R and y in D to $D_{\ell-1}$. Analogously, let D_ℓ contain all arcs between r and B, and for each $y \in Y$ we add all arcs between B and y in D to D_ℓ. Observe that $D_1, D_2, \ldots, D_\ell$ are internally-disjoint S-strong subgraphs in D, so $\kappa_S(D) \geq \ell$.

Conversely, assume that $\kappa_S(D) \geq \ell$ and let $D'_1, D'_2, \ldots, D'_\ell$ be a set of ℓ internally-disjoint S-strong subgraphs in D. At least two of these subgraphs contain no vertex from U (as $|U| = \ell - 2$). Without loss of generality assume that D'_1 and D'_2 do not contain any vertex from U. Let $B' = V(D'_1) \cap X$ and $R' = V(D'_2) \cap X$. Observe that $B' \cap R' = \emptyset$ as D'_1 and D'_2 are internally-disjoint. Every $y \in Y$ has at least one neighbour in B' and R', respectively. Therefore G is 2-colourable (any vertex in G that is not in either R' or B' can be assigned arbitrarily to either B' or R'). This completes the proof. □

In the end of this section, we introduce results for Eulerian digraphs. Using Theorem 2.13, Sun, Gutin and Zhang proved the following result for Eulerian digraphs which means all entries of Table 4.5.

Theorem 4.8 ([77]) *Let $k, \ell \geq 2$ be fixed. For any Eulerian digraph D and $S \subseteq V(D)$ with $|S| = k$, deciding whether $\kappa_S(D) \geq \ell$ is NP-complete.*

Proof We will prove the result by using the NP-completeness of DIRECTED 2-LINKAGE restricted to Eulerian digraphs which was deduced in Theorem 2.13. Let $[D; s_1, t_1, s_2, t_2]$ be an instance of DIRECTED 2-LINKAGE restricted to Eulerian digraphs. Let us first construct a new digraph D' by adding to D vertices x, y, r_1, r_2 and arcs

$$t_1x, xs_1, t_2y, ys_2, xs_2, s_2x, yt_1, t_1y, s_1r_1, r_1t_2, s_2r_2, r_2t_1.$$

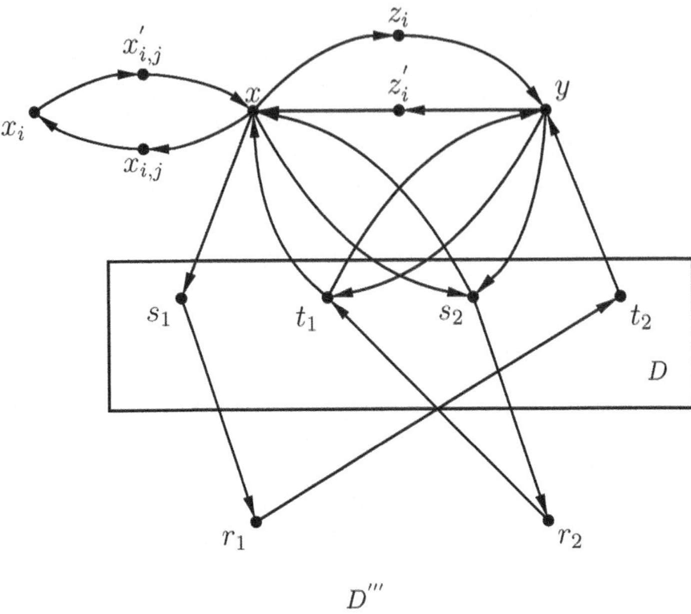

Fig. 4.3 The digraph D'''

Secondly, we add to D' $\ell-2$ copies of the 2-cycle xyx and subdivide the arcs of every copy to avoid parallel arcs, that is, we insert into each arc xy (resp. yx) a new vertex z_i (resp. z'_i) where $i \in [\ell-2]$. Let us denote the new digraph by D''. Note that $D'' = D'$ for $\ell = 2$.

Finally, we add to D'' $k-2$ new vertices $x_1, \dots, x_{k-2}$ and arcs of ℓ 2-cycles xx_ix for each $i \in [k-2]$. Subdivide the new arcs to avoid parallel arcs, that is, we insert each arc xx_i (resp. x_ix) a new vertex $x_{i,j}$ (resp. $x'_{i,j}$) where $j \in [\ell]$. Let us denote the new digraph by D'''. Observe that D''' is Eulerian as D is Eulerian. See Fig. 4.3 for the digraph D'''.

Let $S = \{x, y, x_i \mid i \in [k-2]\}$, $U = \{x_i, x_{i,j}, x'_{i,j} \mid i \in [k-2], j \in [\ell]\}$, $Z = \{z_j \mid j \in [\ell-2]\}$ and $Z' = \{z'_j \mid j \in [\ell-2]\}$. It remains to show that $[D; s_1, t_1, s_2, t_2]$ is a positive instance of DIRECTED 2-LINKAGE restricted to Eulerian digraphs if and only if $\kappa_S(D''') \geq \ell$.

Suppose $[D; s_1, t_1, s_2, t_2]$ is a positive instance of DIRECTED 2-LINKAGE restricted to Eulerian digraphs, that is, there is a pair of vertex-disjoint (s_1, t_1)-path P_1 and (s_2, t_2)-path P_2. Let H_1 be the subdigraph of D''' consisting of the arcs xs_1, t_1x, t_1y, yt_1, the path P_1, and the cycle $x, x_{i,\ell-1}, x_i, x'_{i,\ell-1}, x$ where $i \in [k-2]$. Let H_2 be the subdigraph of D''' consisting of the arcs ys_2, t_2y, s_2x, xs_2 and the path P_2, and the cycle $x, x_{i,\ell}, x_i, x'_{i,\ell}, x$ where $i \in [k-2]$. For $3 \leq j \leq \ell$, let H_j be the subdigraph of D''' consisting of the cycles $x, z_{j-2}, y, z'_{j-2}, x$ and

$x, x_{i,j-2}, x_i, x'_{i,j-2}, x$ where $i \in [k-2]$. Observe that $\{H_i \mid i \in [\ell]\}$ is a family of internally-disjoint S-strong subgraphs, therefore $\kappa_S(D''') \geq \ell$.

If $\kappa_S(D''') \geq \ell$, then there is a set of internally-disjoint S-strong subgraphs, say $\{H_i \mid i \in [\ell]\}$. Observe that $\{H'_i = H_i - U \mid i \in [\ell]\}$ is a set of ℓ internally-disjoint $\{x, y\}$-subgraphs in D''. Since $deg^+_{D''}(x) = deg^-_{D''}(x) = deg^+_{D''}(y) = deg^-_{D''}(y) = \ell$, each H'_i contains precisely one out-neighbour and one in-neighbour of x (resp. y). Therefore, there are two subdigraphs, say H'_1, H'_2, such that $V(H'_i) \cap Z = \emptyset$ for each $i \in [2]$. Let us consider two cases.

Case 1 $V(H'_i) \cap Z' = \emptyset$ for each $i \in [2]$.

We have $V(H'_i) \subseteq V(D')$ for each $i \in [2]$. Since the in-degree of x in D' is 2, we may without loss of generality assume that $t_1 \in V(H'_1)$ and $s_2 \in V(H'_2)$. As y has in-degree 2 in D' and $t_1 \in V(H'_1)$ we must have $t_2 \in V(H'_2)$. As the out-degree of x is 2, we analogously have $s_1 \in V(H'_1)$ (as $s_2 \in V(H'_2)$). Note that now we have both s_i and t_i belong to $V(H'_i)$. Therefore, there must be a path P_i from s_i to t_i in H'_i and by definition of D', P_i will not have vertices outside of D. As H_1 and H_2 are internally-disjoint, the paths are disjoint, so $[D; s_1, t_1, s_2, t_2]$ is a positive instance of DIRECTED 2-LINKAGE restricted to Eulerian digraphs.

Case 2 $V(H'_i) \cap Z' \neq \emptyset$ for some $i \in [2]$.

We will reduce Case 2 to Case 1. We just consider the case that $V(H'_1) \cap Z' \neq \emptyset$ and $V(H'_2) \cap Z' = \emptyset$ since the argument for the remaining case is similar. Let $V(H'_1) \cap Z' = \{z'_1\}$. Observe that there must exist some H'_i, say H'_3, such that $V(H'_3) \cap Z' = \emptyset$. Let P' be a (y, x)-path in H'_3. Then we update H'_1 and H'_3 by exchanging the two paths yz'_1x and P' (note that in this procedure we may need to delete some vertices or arcs to guarantee the strong connectedness of updated H'_1 and H'_3 if necessary, and this will not affect the correctness), and we now also have $V(H'_i) \cap Z' = \emptyset$ for $i \in [2]$ and still make sure that the updated $\{H'_i \mid i \in [\ell]\}$ is a family of ℓ internally-disjoint $\{x, y\}$-subgraphs in D''. □

So far, there is still no result for $\lambda_S(D)$ on semicomplete digraphs, symmetric digraphs and Eulerian digraphs. Therefore, it would be interesting to consider the following question.

Problem 4.4 What is the complexity of deciding whether $\lambda_S(D) \geq \ell$ for integers $k \geq 3$ and $\ell \geq 2$, and a semicomplete digraph (symmetric digraph, Eulerian digraph) D?

4.3 Inapproximability Results on ISSP and ASSP

Recall that in the problem of SET COVER PACKING, the input consists of a bipartite graph $G = (C \cup B, E)$, and the goal is to find a largest collection of pairwise disjoint set covers of B, where a set cover of B is a subset $S \subseteq C$ such that each vertex of B

has a neighbor in C. Feige et al. [25] proved the following inapproximability result on the SET COVER PACKING problem.

Theorem 4.9 ([25]) *Unless P=NP, there is no $o(\log n)$-approximation algorithm for* SET COVER PACKING, *where n is the order of G.*

Sun, Gutin and Zhang got two inapproximability results on ISSP and ASSP by reductions from the SET COVER PACKING problem.

Theorem 4.10 ([77]) *The following assertions hold:*

(a) Unless P=NP, there is no $o(\log n)$-approximation algorithm for ISSP, *even restricted to the case that D is a symmetric digraph and S is independent in D, where n is the order of D.*
(b) Unless P=NP, there is no $o(\log n)$-approximation algorithm for ASSP, *even restricted to the case that S is independent in D, where n is the order of D.*

Proof **Part (a)** Let $G = (C \cup B, E)$ be an instance of SET COVER PACKING. We construct an instance $[D; S]$ of ISSP by setting

$$V(D) = \{x\} \cup C \cup B,$$

$$A(D) = \{xu, ux \mid u \in C\} \cup \{uv, vu \mid u \text{ and } v \text{ are adjacent in } G\}$$

and

$$S = \{x\} \cup B.$$

If $\{C_i \subseteq C \mid i \in [\ell]\}$ is a set cover packing, then the subdigraphs in D induced by the vertex sets $\{x\} \cup C_i \cup B$ form a set of ℓ internally-disjoint S-strong subgraphs in D.

Conversely, let $\{D_i \mid i \in [\ell]\}$ be a set of ℓ internally-disjoint S-strong subgraphs in D. Since B is an independent set in D, for each D_i, there is a set $C_i \subseteq C$ of vertices satisfying the following: every vertex in B has a neighbor in C_i such that it can reach the vertex x. Observe that these sets C_i are pairwise disjoint and form a set cover packing of cardinality ℓ. Note that D is symmetric and S is an independent set of D. This completes the proof of (a) by Theorem 4.9.

Part (b) We construct an instance $[D'; S']$ of ASSP from $[D; S]$ constructed in part (i) with $V(D') = \{x\} \cup B \cup \{u^+, u^- \mid u \in C\}$ and $S' = S = \{x\} \cup B$ such that:

(1) $u^-u^+ \in A(D')$ for each $u \in C$;
(2) $vu^- \in A(D')$ if $vu \in A(D)$, $v \in S'$, $u \in C$;
(3) $u^+v \in A(D')$ if $uv \in A(D)$, $v \in S'$, $u \in C$.

If $\{C_i' \subseteq C \mid i \in [\ell]\}$ is a set cover packing, then the subdigraph in D' induced by the vertex set $\{x\} \cup \{u^-, u^+ \mid u \in C_i'\} \cup B$ $(i \in [\ell])$ forms a set of ℓ arc-disjoint S'-strong subgraphs in D'.

Now let $\{D'_i \mid i \in [\ell]\}$ be a set of ℓ arc-disjoint S'-strong subgraphs in D'. In each D'_i, since B is an independent set in D', each vertex $v \in B$ has to pass through an arc of type u^-u^+ to reach x for some $u \in C$. Hence, in G there is a set $C'_i \subseteq C$ of vertices such that every vertex in B has a neighbor in C'_i. Furthermore, since the strong subgraphs D'_i are pairwise arc-disjoint, the sets C'_i $(i \in [\ell])$ are pairwise disjoint and form a set cover packing of cardinality ℓ. Note that S' is an independent set of D'. This completes the proof of (b) by Theorem 4.9. □

4.4 Related Topics

4.4.1 Kriesell Conjecture and Its Extension in Strong Subgraph Packing Problem

Let G be a connected graph with $S \subseteq V(G)$. We say that a set of edges C of G an *S-Steiner-cut* if there are at least two components of $G - C$ which contain vertices of S. Similarly, let D be a strong digraph and $S \subseteq V(D)$, we say that a set of arcs C of D an *S-strong subgraph-cut* if there are at least two strong components of $D - C$ which contain vertices of S.

Kriesell posed the following well-known conjecture which concerns an approximate min-max relation between the size of an S-Steiner-cut and the number of edge-disjoint S-Steiner trees. This type of problem is also called the Erdős-Pósa type problem.

Conjecture 4.1 ([47]) Let G be a graph and $S \subseteq V(G)$ with $|S| \geq 2$. If every S-Steiner-cut in G has size at least 2ℓ, then G contains ℓ pairwise edge-disjoint S-Steiner trees.

Lau [52] proved that the conjecture holds if every S-Steiner-cut in G has size at least 26ℓ. West and Wu [84] improved the bound significantly by showing that the conjecture still holds if 26ℓ is replaced by 6.5ℓ. So far the best bound $5\ell + 4$ was obtained by DeVos, McDonald and Pivotto as follows.

Theorem 4.11 ([22]) *Let G be a graph and $S \subseteq V(G)$ with $|S| \geq 2$. If every S-Steiner-cut in G has size at least $5\ell + 4$, then G contains ℓ pairwise edge-disjoint S-Steiner trees.*

Similar to Theorem 4.11, it is natural to study an approximate min-max relation between the size of minimum S-strong subgraph-cut and the maximum number of arc-disjoint S-strong subgraphs in a digraph D. Here is an interesting problem which is analogous to Conjecture 4.1.

Problem 4.5 ([77]) Let D be a digraph and $S \subseteq V(D)$ with $|S| \geq 2$. Find a function $f(\ell)$ such that the following holds: If every S-strong subgraph-cut in D has size at least $f(\ell)$, then D contains ℓ pairwise arc-disjoint S-strong subgraphs.

Note that there is a linear function $f(\ell)$ for a strong symmetric digraph: Let $D = \overleftrightarrow{G}$ be a strong symmetric digraph and $S \subseteq V(D)$. If every S-strong subgraph-cut in D has size at least $10\ell + 8$, then D contains ℓ pairwise arc-disjoint S-strong subgraphs. The argument is as follows: Let c_1 and c_2 be the sizes of the minimum S-Steiner-cut in G and the minimum S-strong subgraph-cut in D, respectively. We deduce that $c_1 \geq \frac{c_2}{2}$. Indeed, let $C_1 = \{e_i \mid i \in [c_1]\}$ be the minimum S-Steiner-cut in G. Let $C_1' = \{a_i, a_i' \mid i \in [c_1]\}$, where a_i, a_i' be the two arcs in D corresponding to the edge e_i. It can be checked that C_1' is an S-strong subgraph-cut of D. Hence, $c_2 \leq |C_1'| = 2c_1$. The assumption means that $c_2 \geq 10\ell + 8$ and so $c_1 \geq \frac{c_2}{2} \geq 5\ell + 4$. By Theorem 4.11, G contains ℓ pairwise edge-disjoint S-Steiner trees. For each S-Steiner tree, we can obtain an S-strong subgraph in D by replacing each edge of this tree with the corresponding arcs of both directions in D. Observe that we now obtain ℓ pairwise arc-disjoint S-strong subgraphs.

4.4.2 Strong Subgraph Connectivity

Related to strong subgraph packing problems, there is another new connectivity of digraphs, called strong subgraph connectivity, including strong subgraph k-connectivity and strong subgraph k-arc-connectivity. The *strong subgraph k-connectivity* [76] is defined as

$$\kappa_k'(D) = \min\{\kappa_S(D) \mid S \subseteq V(D), |S| = k\}.$$

Similarly, the *strong subgraph k-arc-connectivity* [68] is defined as

$$\lambda_k'(D) = \min\{\lambda_S(D) \mid S \subseteq V(D), |S| = k\}.$$

The strong subgraph connectivity is not only a natural extension of tree connectivity [38, 53, 54] of undirected graphs to directed graphs, but also could be seen as a generalization of classical connectivity of undirected graphs by the fact that $\kappa_2'(\overleftrightarrow{G}) = \kappa(G)$ [76] and $\lambda_2'(\overleftrightarrow{G}) = \lambda(G)$ [68]. For more information on this topic, the reader can see [23, 68, 69, 76, 77] and a recent survey [67].

Chapter 5
Strong Arc Decomposition Problem

Abstract Recall that it is NP-complete to decide whether a digraph has a strong arc decomposition (Theorem 1.2), therefore it is natural to consider this problem on some digraph classes. In this chapter, we introduce most of known results on semicomplete digraphs, locally semicomplete digraphs, digraph compositions and digraph products. We also introduce Kelly conjecture and Bang-Jensen–Yeo conjecture which are highly related to the strong arc decomposition problem.

5.1 Semicomplete Digraphs

Clearly, every digraph with a strong arc decomposition is 2-arc-strong. Bang-Jensen and Yeo characterized semicomplete digraphs with a strong arc decomposition in Theorem 5.2. In order to prove this theorem, they gave several preliminary results as follows.

Lemma 5.1 ([12]) *Every 2-arc-strong semicomplete digraph H has three distinct vertices q_1, q_2, q_3 such that $H - q_i$ is strong for $i \in [3]$.*

Lemma 5.2 ([12]) *Let D be a k-arc-strong digraph, and let C be a cycle in D. Then the digraph obtained by reversing all arcs in the cycle C is also k-arc-strong.*

Lemma 5.3 ([12]) *Let D be a k-arc-strong semicomplete digraph, and let $x \in V(D)$, have the property that $d^+(w) = k$, for all $w \in N^+(x)$. If $\delta^0(D - x) \geq k$, then $D - x$ is k-arc-strong.*

Theorem 5.1 ([12]) *Let $k \geq 1$ and let D be a k-arc-strong semicomplete digraph such that there a set $S \subset V(D)$, with $2 \leq |S| \leq |V(D)| - 2$ and $d^+(S) = k$. There exist k arc-disjoint spanning strong subgraphs of D except if $D = S_4$.*

By Theorems 1.1 and 5.1, Lemmas 5.1–5.3, Bang-Jensen and Yeo obtained the following characterization for semicomplete digraphs with a strong arc decomposition.

Theorem 5.2 ([12]) *A 2-arc-strong semicomplete digraph D has a strong arc decomposition if and only if D is not isomorphic to S_4, where S_4 is obtained from*

Y. Sun, *Steiner Type Packing Problems in Digraphs*, SpringerBriefs in Mathematics,
https://doi.org/10.1007/978-981-95-8743-8_5

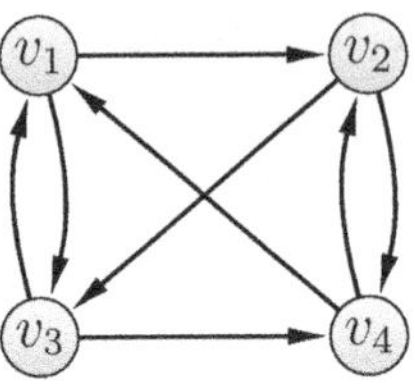

Fig. 5.1 The digraph S_4

the complete digraph with four vertices by deleting a cycle of length 4 (see Fig. 5.1). Furthermore, a strong arc decomposition of D can be obtained in polynomial time when it exists.

5.2 Locally Semicomplete Digraphs

Bang-Jensen and Huang [9] studied the strong arc decomposition in a larger digraph classes, locally semicomplete digraphs, as shown in Theorem 5.6. In order to prove this theorem, we need some preliminary results.

We begin with the structure of non-strong locally semicomplete digraphs.

Lemma 5.4 ([14]) *Every non-strong locally semicomplete digraph D has a unique round decomposition $R[D_1, D_2, \ldots, D_p]$, where $D_1, D_2, \ldots, D_p$ is the unique acyclic ordering of the strong components of D and R is a round local tournament containing no cycle.*

For convenience, we call the unique acyclic ordering $D_1, D_2, \ldots, D_p$ in Lemma 5.4 the *strong decomposition* of D with the *initial component* D_1 and the *terminal component* D_p.

There is another kind of decomposition of non-strong locally semicomplete digraphs is also useful in the proofs.

Theorem 5.3 ([35]) *Let D be a non-strong locally semicomplete digraph and let $D_1, D_2, \ldots, D_p$ be the strong decomposition of D. Then D can be decomposed into $r \geq 2$ subdigraphs $D'_1, D'_2, \ldots, D'_r$ as follows:*

$$D'_1 = D_p, \psi_1 = p,$$

$$\psi_{i+1} = \min\{j \mid N^+(D_j) \cap V(D'_i) \neq \emptyset\},$$

$$D'_{i+1} = D[V(D_{\psi_{i+1}}) \cup V(D_{\psi_{i+1}+1}) \cup \cdots \cup V(D_{\psi_i - 1})].$$

The subdigraphs $D'_1, D'_2, \ldots, D'_r$ satisfies the properties below:

(a) D'_i consists of some strong components of D and is semicomplete for $i \in [r]$;

(b) D'_{i+1} *dominates the initial component of* D'_i *and there exists no arc from* D'_i *to* D'_{i+1} *for* $i \in [r-1]$*;*
(c) *if* $r \geq 3$*, then there is no arc between* D'_i *and* D'_j *for* i, j *satisfying* $|j-i| \geq 2$*.*

The unique sequence $D'_1, D'_2, \ldots, D'_r$ defined in Theorem 5.3 will be referred to as the *semicomplete decomposition* of D.

We now turn to the structure of strong locally semicomplete digraphs.

Lemma 5.5 ([14]) *Let* D *be a strong locally semicomplete digraph which is not semicomplete. Then* D *is not round decomposable if and only if the following conditions are satisfied:*

(a) *There is a minimal separating set* S *such that* $D - S$ *is not semicomplete and for each such* S*,* $D[S]$ *is semicomplete and the semicomplete decomposition of* $D - S$ *has exactly three components* D'_1, D'_2, D'_3*;*
(b) *Furthermore, for each such* S*, there are integers* $\alpha, \beta, \mu, \upsilon$ *with* $\psi_2 \leq \alpha \leq \beta \leq p-1$ *and* $p+1 \leq \mu \leq \upsilon \leq p+q$ *such that*

$$N^-(D_\alpha) \cap V(D_\mu) \neq \emptyset, N^+(D_\alpha) \cap V(D_\upsilon) \neq \emptyset,$$

or

$$N^-(D_\mu) \cap V(D_\alpha) \neq \emptyset, N^+(D_\mu) \cap V(D_\beta) \neq \emptyset,$$

where $D_1, D_2, \ldots, D_p$ *and* $D_{p+1}, D_{p+2}, \ldots, D_{p+q}$ *are the strong decomposition of* $D - S$ *and* $D[S]$*, respectively, and* D_{ψ_2} *is the initial component of* D'_2*.*

We can now give a full classification of locally semicomplete digraphs.

Theorem 5.4 ([14]) *Let* D *be a locally semicomplete digraph. Then exactly one of the following possibilities holds.*

(a) D *is round decomposable with a unique round decomposition* $R[D_1, D_2, \ldots, D_\alpha]$*, where* R *is a round local tournament on* $\alpha \geq 2$ *vertices and* D_i *is a strong semicomplete digraph for* $i \in [\alpha]$*;*
(b) D *is not round decomposable and not semicomplete and it has the structure as described in Lemma 5.5;*
(c) D *is a semicomplete digraph which is not round decomposable.*

The following seven results are frequently used in the proof of Theorem 5.6.

Lemma 5.6 ([14]) *Let* D *be a strong non-round decomposable locally semicomplete digraph and let* S *be a minimal separating set of* D *such that* $D - S$ *is not semicomplete. Let* $D_1, D_2, \ldots, D_p$ *be the strong decomposition of* $D - S$ *and* $D_{p+1}, D_{p+2}, \ldots, D_{p+q}$ *be the strong decomposition of* $D[S]$*. Then the following holds:*

- $D_p \Rightarrow S \Rightarrow D_1$.
- *Suppose that there is an arc* $s \to v$ *from* S *to* D'_2 *with* $s \in V(D_i)$ *and* $v \in V(D_j)$. *Then*

$$D_i \cup D_{i+1} \cup \cdots \cup D_{p+q} \Rightarrow D'_3 \Rightarrow D_{\psi_2} \cup \cdots \cup D_j.$$

- $D_{p+q} \Rightarrow D'_3$ *and* $D_f \Rightarrow D_{f+1}$ *for* $f \in [p+q]$ *where indices are modulo* $p+q$.

Lemma 5.7 ([9]) *Let* $D = \overrightarrow{C}_4[\{u\}, X, \{v\}, Y]$ *where* $\min\{|X|, |Y|\} \geq 2$. *If the minimum semi-degree of* $D[X \cup Y]$ *is at least 1 then* D *is 2-arc-strong.*

Theorem 5.5 ([2]) *Every strong locally semicomplete digraph* D *which is not a cycle has a vertex* w *such that* $D - w$ *is also strong.*

Lemma 5.8 ([9]) *A 2-arc-strong round local tournament* D *has a strong arc decomposition if and only if* D *is not the second power of an even cycle.*

Lemma 5.9 ([9]) *Let* D *be a locally semicomplete digraph which is obtained from* $\overrightarrow{C}^2_{2k}$ *by adding a new vertex* x, *at least two arcs from* x *to* $V(\overrightarrow{C}^2_{2k})$ *and at least two arcs from* $V(\overrightarrow{C}^2_{2k})$ *to* x. *Then* D *has a strong arc decomposition.*

Lemma 5.10 ([9]) *Let* $D = R[D_1, D_2, \ldots, D_r]$, $r \geq 3$ *be a 2-arc-strong round decomposable locally semicomplete digraph for which at least one* D_i *has more than one vertex. Then* D *contains a Hamiltonian cycle* C *such that* $D - A(C)$ *is strong. Furthermore we can find* C *in polynomial time.*

Lemma 5.11 ([9]) *Let* $D = (V, A)$ *be a locally semicomplete digraph on at least 5 vertices which is 2-arc-strong but not 2-strong. Then* D *has a strong arc decomposition and we can find such a decomposition in polynomial time.*

Based on the structural characterization of locally semicomplete digraphs given by Theorem 5.4, Bang-Jensen and Huang[9] obtained the following characterization of locally semicomplete digraphs with a strong arc decomposition. Their proof is quite non-trivial and long, so we omit the details.

Theorem 5.6 ([9]) *A 2-arc-strong locally semicomplete digraph* D *has a strong arc decomposition if and only if* D *is not the second power of an even cycle.*

Note that the proof of Theorem 5.6 can be turned into polynomial algorithms for finding a strong arc decomposition in a locally semicomplete digraph which is not the second power of an even cycle.

Corollary 5.1 ([9]) *There exists a polynomial algorithm for finding a strong arc decomposition of a given 2-arc-strong locally semicomplete digraph* D *which is not the second power of an even cycle.*

Since no second power of a cycle is 3-arc-strong, Theorem 5.6 implies the following.

Corollary 5.2 ([9]) *Every 3-arc-strong locally semicomplete digraph has a strong arc decomposition.*

Theorem 5.6 also implies the following result.

Corollary 5.3 ([9]) *Every 2-arc-strong locally semicomplete digraph $D = (V, A)$ contains arc-disjoint in- and out-branchings B_u^-, B_v^+ for every choice of $u, v \in V$.*

Bang-Jensen and Huang posed a conjecture which could be seen as a natural extension of Theorem 5.6.

Conjecture 5.1 ([9]) A k-arc-strong locally semicomplete digraph D can be decomposed into k arc-disjoint spanning strong subgraphs if and only if D is not $\overrightarrow{C}_n^k$, where n is divisible by some $i \in [k]$.

The above conjecture implies the following:

Conjecture 5.2 ([9]) For every natural number k there exists a natural number $f(k)$ such that every k-arc-strong locally semicomplete digraph D with $\delta^0(D) \geq f(k)$ has an arc decomposition $A(D) = A_1 \cup A_2 \cup \cdots \cup A_k$ such that each of the spanning subdigraphs $D_i = (V(D), A_i)$ is strong for $i \in [k]$.

Thomassen [78] conjectured that every 3-strong tournament has two arc-disjoint Hamiltonian cycles. Bang-Jensen and Huang extended this to local tournaments.

Conjecture 5.3 ([9]) Every 3-strong local tournament has two arc-disjoint Hamiltonian cycles.

5.3 Digraph Compositions

Sun et al. [75] considered strong arc decompositions on digraph compositions and products. Let us start from a simple observation, which will be useful in the proofs of the theorems of this section.

Lemma 5.12 ([75]) *Let D be a digraph on t vertices ($t \geq 2$) and let $H_1', \ldots, H_t'$ be digraphs with no arcs. If an induced subdigraph Q^* of $Q' = D[H_1', \ldots, H_t']$ with at least one vertex in each H_i', $i \in [t]$ has a strong arc decomposition, then so have Q'.*

Proof Let $\{u_{i,1}, \ldots, u_{i,n_i}\}$ be the set of vertices of H_i' for every $i \in [t]$. For every $i \in [t]$, let $H_i^{(m_i)}$ be the subdigraph of H_i' induced by $\{u_{i,1}, \ldots, u_{i,m_i}\}$, where $1 \leq m_i \leq n_i$. Without loss of generality, let $Q^* = D[H_1^{(m_1)}, \ldots, H_t^{(m_t)}]$ and let Q^* have a decomposition into arc-disjoint strong spanning subdigraphs Q_1^*, Q_2^*. To extend this decomposition to Q', for every i, j, where $i \in [t]$ and $j \in \{1, 2\}$, add to Q_j^* the vertices $u_{i,m_i+1}, \ldots, u_{i,n_i}$ and let them have the same in- and out-neighbours as $u_{i,1}$. (This way the inserted vertices will keep Q_1^* and Q_2^* strongly connected.) □

The following theorem gives sufficient conditions for a digraph composition to have a strong arc decomposition.

Theorem 5.7 ([75]) *Let T be a digraph with vertices $u_1, \dots, u_t$ $(t \geq 2)$ and let $H_1, \dots, H_t$ be digraphs. Let the vertex set of H_i be $\{u_{i,j_i} \mid i \in [t], j_i \in [n_i]\}$ for every $i \in [t]$. Then $Q = T[H_1, \dots, H_t]$ has a strong arc decomposition if at least one of the following conditions holds:*

(a) T is a 2-arc-strong semicomplete digraph and $H_1, \dots, H_t$ are arbitrary digraphs, but Q is not isomorphic to S_4 (see Fig. 5.1);
(b) T has a Hamiltonian cycle and one of the following conditions holds:

- *t is even and $n_i \geq 2$ for every $i \in [t]$;*
- *t is odd, $n_i \geq 2$ for every $i \in [t]$ and at least two distinct subdigraphs H_i have arcs;*
- *t is odd and $n_i \geq 3$ for every $i \in [t]$ apart from one i for which $n_i \geq 2$.*

(c) T and all H_i are strong digraphs with at least two vertices.

Proof For every $i \in [t]$, let H_i' be the digraph obtained from H_i by deleting all arcs. Let $Q' = T[H_1', \dots, H_t']$. We will prove parts of the theorem one by one.
Part (a) If T is not isomorphic to S_4 then we are done by Theorem 5.2 and Lemma 5.12. Now assume that T is isomorphic to S_4, but Q is not isomorphic to S_4. Let the vertices of T be u_1, u_2, u_3, u_4 and its arcs

$$u_1u_2, u_2u_1, u_3u_4, u_4u_3, u_1u_4, u_2u_3, u_4u_2, u_3u_1.$$

Since Q is not isomorphic to S_4, at least one of H_1, H_2, H_3, H_4 has at least two vertices. Without loss of generality, let H_1 have at least two vertices. Consider the subdigraph Q^* of Q' induced by $\{u_{1,1}, u_{1,2}, u_{2,1}, u_{3,1}, u_{4,1}\}$. Then Q^* has two arc-disjoint spanning strong subgraphs: Q_1^* with arcs

$$\{u_{1,1}u_{2,1}, u_{2,1}u_{1,2}, u_{1,2}u_{4,1}, u_{4,1}u_{3,1}, u_{3,1}u_{1,1}\}$$

and Q_2^* with arcs

$$\{u_{2,1}u_{1,1}, u_{1,1}u_{4,1}, u_{4,1}u_{2,1}, u_{2,1}u_{3,1}, u_{3,1}u_{1,2}, u_{1,2}u_{2,1}\}.$$

It remains to apply Lemma 5.12 to obtain a strong arc decomposition of Q' and thus of Q.
Part (b) Without loss of generality, assume that $u_1u_2 \dots u_tu_1$ is a Hamiltonian cycle of T. Let $U = \bigcup_{i=1}^{t}\{u_{i,1}, u_{i,2}\}$.

Case 1 t is even and $n_i \geq 2$ for every $i \in [t]$.

The following arc sets induce arc-disjoint spanning strong subgraphs Q_1^*, Q_2^* of $Q'[U]$:

$$\{u_{i,j}u_{i+1,j} \mid i \in [t-1], j \in [2]\} \cup \{u_{t,1}u_{1,2}, u_{t,2}u_{1,1}\}; \tag{5.1}$$

$$\{u_{i,j}u_{i+1,j'} \mid i \in [t-1], j \in [2]\} \cup \{u_{t,1}u_{1,1}, u_{t,2}u_{1,2}\}, \tag{5.2}$$

where $j' = j + 1 \pmod 2$.

It remains to apply Lemma 5.12 to obtain a strong arc decomposition of Q' and thus of Q.

Case 2 t is odd, $n_i \geq 2$ for every $i \in [t]$ and at least two distinct subdigraphs H_i have arcs.

Let e_p, e_q be arcs in two distinct subdigraphs H_p and H_q. We may assume that both end-vertices of e_p and e_q are in U. Observe that while Q_1^* (with arcs listed in (5.1)) is strong, Q_2^* (with arcs listed in (5.2)) forms two arc-disjoint cycles C and Z. We may assume that the tail (head) of e_p (e_q) is in C and the head (tail) of e_p (e_q) is in Z (otherwise, relabel vertices in $\{u_{p,1}, u_{p,2}\}$ and/or $\{u_{q,1}, u_{q,2}\}$). Thus, adding e_p and e_q to Q_2^* makes it strong. To obtain two arc-disjoint spanning strong subgraphs of Q from Q_1^*, Q_2^*, let every vertex $u_{i,j}$ for $j \geq 3$ and $1 \leq i \leq t$ have the same out- and in-neighbours as $u_{i,1}$ in Q'.

Case 3 t is odd and $n_i \geq 3$ for every $i \in [t]$ apart from one i for which $n_i \geq 2$.

Without loss of generality, assume that $n_1 \geq 2$ and $n_i \geq 3$ for all $i \in \{2, 3, \ldots, t\}$.

First we consider the subcase in which $t = 3$, $n_1 = 2$, and $n_2 = n_3 = 3$. Then Q' has two arc-disjoint spanning subdigraphs Q_1^* and Q_2^* with arc sets

$$\{u_{1,1}u_{2,1}, u_{3,1}u_{1,1}, u_{1,2}u_{2,2}, u_{1,2}u_{2,3}, u_{3,2}u_{1,2}, u_{3,3}u_{1,2}, u_{2,1}u_{3,2}, u_{2,2}u_{3,1}, u_{2,3}u_{3,3}\},$$

$$\{u_{1,1}u_{2,2}, u_{1,1}u_{2,3}, u_{3,2}u_{1,1}, u_{3,3}u_{1,1}, u_{1,2}u_{2,1}, u_{3,1}u_{1,2}, u_{2,1}u_{3,3}, u_{2,2}u_{3,2}, u_{2,3}u_{3,1}\},$$

respectively. Observe that Q_1^* and Q_2^* are strong since they contain the closed walks through all vertices, respectively:

$$u_{1,2}u_{2,2}u_{3,1}u_{1,1}u_{2,1}u_{3,2}u_{1,2}u_{2,3}u_{3,3}u_{1,2}; \; u_{1,1}u_{2,2}u_{3,2}u_{1,1}u_{2,3}u_{3,1}u_{1,2}u_{2,1}u_{3,3}u_{1,1}.$$

Now we extend the previous subcase to that in which $n_1 = 2$ and $n_i = 3$ for all $i \in \{2, 3, \ldots, t\}$. First replace index 3 in every vertex of the form $u_{3,i}$ by t in the two arc sets of the previous subcase. Then replace every arc of the form $u_{2,i}u_{t,j}$ in Q_1^* by the path $u_{2,i}u_{3,i} \ldots u_{t-1,i}u_{t,j}$. In Q_2^*, we replace $u_{2,1}u_{t,3}$ by the path $u_{2,1}u_{3,2}u_{4,1}u_{5,2} \ldots u_{t-1,1}u_{t,3}$, replace $u_{2,2}u_{t,2}$ by the path $u_{2,2}u_{3,1}u_{4,2}u_{5,1} \ldots u_{t-1,2}u_{t,2}$, replace $u_{2,3}u_{t,1}$ by the path $u_{2,3}u_{3,2}u_{4,3}u_{5,2} \ldots u_{t-1,3}u_{t,1}$, and finally add the path $u_{2,2}u_{3,3}u_{4,2}u_{5,3} \ldots u_{t-1,2}$.

Finally, we extend the previous subcase to the general one using Lemma 5.12.

Part (c) For $j \in [2]$, let T_j be the subdigraph of Q induced by vertex set $\{u_{i,j} \mid i \in [t]\}$. Clearly, $T_1 \cong T_2 \cong T$ and T_1 and T_2 are strong.

Let Q_1 be the spanning subdigraph of Q with arc set $A(Q_1) = A(T_1) \cup (\bigcup_{i=1}^{t} A(H_i))$. Observe that Q_1 is strong since T_1 and each H_i are strong, and T_1 has a common vertex with each H_i, where $i \in [t]$.

Let Q_2 be the spanning subdigraph of Q with arc set $A(Q_2) = A(Q)\setminus A(Q_1)$. To see that Q_2 is strong, we only need to find a strong subgraph in Q_2 which contains x and y for each pair of distinct vertices x and y in Q_2. We will consider two cases.

Case 1 $x \in V(T_1)$.

Without loss of generality, we assume that $x = u_{1,1}$ and $y \in \{u_{1,2}, u_{2,1}, u_{2,2}\}$. We first consider the subcase that $y = u_{2,1}$. Observe that there is at least one arc entering and one arc leaving $u_{1,2}$ ($u_{2,2}$) in T_2, and so there are two arcs, say a and b (c and d), with opposite directions between x (y) and T_2 in Q_2. Then by adding the arcs a, b, c, d, and the vertices x, y to T_2, we obtain a strong subgraph T_2' of Q_2 which contains both x and y, as desired. For the case that $y \in \{u_{1,2}, u_{2,2}\}$, we just add the arcs a, b, and the vertex x to T_2, and then obtain a strong subgraph T_2'' of Q_2 which contains both x and y.

Case 2 $x \notin V(T_1)$.

Without loss of generality, we assume that $x = u_{1,2}$ and $y \in \{u_{1,1}, u_{2,1}, u_{1,3}, u_{2,2}, u_{2,3}\}$ (if $u_{1,3}$ and $u_{2,3}$ exist). By Case 1 and the fact that $T_2 \cong T$ is strong, we are done if $y \in \{u_{1,1}, u_{2,2}\}$. For the case that $y = u_{2,1}$, by adding the arcs c, d and the vertex y to T_2, we can obtain a strong subgraph T_2''' of Q_2 which contains both x and y. With a similar argument, we can get the desired strong subgraph for the case that $y \in \{u_{1,3}, u_{2,3}\}$.

Hence, we complete the argument and conclude that Q has a strong arc decomposition. □

Sun et al. [75] used Theorem 5.7 to prove the following characterization for semicomplete compositions $T[H_1, \ldots, H_t]$ when each H_i has at least two vertices.

Theorem 5.8 ([75]) *Let $Q = T[H_1, \ldots, H_t]$ be a strong semicomplete composition such that each H_i has at least two vertices for $i \in [t]$. Then Q has a strong arc decomposition if and only if Q is not isomorphic to one of the following three digraphs: $\overrightarrow{C}_3[\overline{K}_2, \overline{K}_2, \overline{K}_2]$, $\overrightarrow{C}_3[\overrightarrow{P}_2, \overline{K}_2, \overline{K}_2]$, $\overrightarrow{C}_3[\overline{K}_2, \overline{K}_2, \overline{K}_3]$.*

Proof Let us first prove the "only if" part of the theorem, i.e. $\overrightarrow{C}_3[\overline{K}_2, \overline{K}_2, \overline{K}_2]$, $\overrightarrow{C}_3[\overrightarrow{P}_2, \overline{K}_2, \overline{K}_2]$ and $\overrightarrow{C}_3[\overline{K}_2, \overline{K}_2, \overline{K}_3]$ do not have strong arc decompositions. By Lemma 5.12, it suffices to show that neither $\overrightarrow{C}_3[\overrightarrow{P}_2, \overline{K}_2, \overline{K}_2]$ nor $\overrightarrow{C}_3[\overline{K}_2, \overline{K}_2, \overline{K}_3]$ has a strong arc decomposition. The proof is by reductio ad absurdum.

Suppose that $Q = \overrightarrow{C}_3[\overrightarrow{P}_2, \overline{K}_2, \overline{K}_2]$ has a decomposition into two spanning strong subgraphs D_1, D_2. Since Q has 13 arcs, without loss of generality, we may assume that D_1 is a Hamiltonian cycle of Q. Since the arc of H_1 cannot be in a Hamiltonian cycle of Q, without loss of generality, let $D_1 = u_{1,1}u_{2,1}u_{3,1}u_{1,2}u_{2,2}u_{3,2}u_{1,1}$. Then the remaining arcs of Q form two disjoint cycles

$u_{1,1}u_{2,2}u_{3,1}u_{1,1}$ and $u_{1,2}u_{2,1}u_{3,2}u_{1,2}$ and a single arc between them, a contradiction to the assumption that D_2 is strong.

Suppose that $Q = \overrightarrow{C}_3[\overline{K}_2, \overline{K}_2, \overline{K}_3]$ has a decomposition into two spanning strong subgraphs D_1, D_2. Since Q has 16 arcs and has no Hamiltonian cycle, each of D_1, D_2 has 8 arcs. Since Q has only cycles of lengths 3 and 6 and D_1 is strong, without loss of generality, we may assume that D_1 consists of a cycle $u_{1,1}u_{2,1}u_{3,1}u_{1,2}u_{2,2}u_{3,2}u_{1,1}$ and a path $u_{2,1}u_{3,3}u_{1,1}$. Then D_2 consists of two cycles $u_{1,1}u_{2,2}u_{3,1}u_{1,1}$ and $u_{1,2}u_{2,1}u_{3,2}u_{1,2}$ and a path $u_{2,2}u_{3,3}u_{1,2}$. Observe that D_2 is not strong, a contradiction.

Now we will show the "if" part of the theorem by reductio ad absurdum as well. Assume that Q is not isomorphic to either of the three digraphs, but has no strong arc decomposition.

By Camion's Theorem (see [18] or Theorem 2.2.6 of [6]), T has a Hamiltonian cycle $C = u_1u_2 \dots u_tu_1$. Thus, Conditions (b) of Theorem 5.7 are applicable. By the conditions, t must be odd and for at least two distinct indexes $p, q \in [t]$, we have $n_p = n_q = 2$.

Suppose $t \geq 5$. Then there will be arcs between H_i and H_{i+2} in Q for every $i \in [t-2]$. Recall Case 2 of Part (b) of the proof of Theorem 5.7. The arcs between H_i and H_{i+2} can be used to make D_2 strong instead of arcs e_p and e_q used in Case 2 of Part (b) of the proof of Theorem 5.7. Thus, Q has a strong arc decomposition, a contradiction. Hence, $t = 3$ and, without loss of generality, $n_1 = n_2 = 2$ and $n_3 \geq 2$.

Suppose that T has opposite arcs. One of these arcs will not be on the Hamiltonian cycle C of T and will correspond to four or more arcs in Q. Now recall Case 2 of Part (b) of the proof of Theorem 5.7. Two of the above-mentioned arcs can be used to make D_2 strong instead of arcs e_p and e_q used in Case 2 of Part (b) of the proof of Theorem 5.7. Thus, Q has a strong arc decomposition, a contradiction. Hence, $T = \overrightarrow{C}_3$.

Suppose that $n_3 \geq 4$. To get a contradiction, by Lemma 5.12 it suffices to show that $Q = \overrightarrow{C}_3[\overline{K}_2, \overline{K}_2, \overline{K}_4]$ has a decomposition into two spanning strong subgraphs D_1, D_2, where D_1 consists of a cycle $u_{1,1}u_{2,1}u_{3,1}u_{1,2}u_{2,2}u_{3,2}u_{1,1}$ and two paths $u_{2,1}u_{3,4}u_{1,1}$ and $u_{2,2}u_{3,3}u_{1,2}$ and D_2 consists of two cycles $u_{1,1}u_{2,2}u_{3,1}u_{1,1}$ and $u_{1,2}u_{2,1}u_{3,2}u_{1,2}$ and two paths $u_{2,1}u_{3,3}u_{1,1}$ and $u_{2,2}u_{3,4}u_{1,2}$. Thus, $n_3 \leq 3$.

Now consider the case of $n_1 = n_2 = 2$ and $n_3 = 3$. Since Q is not isomorphic to $\overrightarrow{C}_3[\overline{K}_2, \overline{K}_2, \overline{K}_3]$, it has an arc in either H_1 or H_2 or H_3, and by Conditions (b) of Theorem 5.7, only one of H_1, H_2, H_3 has an arc a. Without loss of generality, assume that if H_1 has an arc then $a = u_{1,2}u_{1,1}$, if H_2 has an arc then $a = u_{2,1}u_{2,2}$ and if H_3 has an arc then $a = u_{3,2}u_{3,1}$. Then Q has a decomposition into two spanning subdigraphs D_1, D_2, where D_1 consists of a cycle $u_{1,1}u_{2,1}u_{3,1}u_{1,2}u_{2,2}u_{3,2}u_{1,1}$ and a path $u_{2,1}u_{3,3}u_{1,1}$ and D_2 consists of two cycles $u_{1,1}u_{2,2}u_{3,1}u_{1,1}$ and $u_{1,2}u_{2,1}u_{3,2}u_{1,2}$, a path $u_{2,2}u_{3,3}u_{1,2}$ and arc a. Observe that both D_1 and D_2 are strong, a contradiction.

It remains to consider the case of $n_1 = n_2 = n_3 = 2$. Since Q is not isomorphic to $\overrightarrow{C}_3[\overline{K}_2, \overline{K}_2, \overline{K}_2]$, at least one of H_1, H_2 and H_3 has an arc. By Conditions (b)

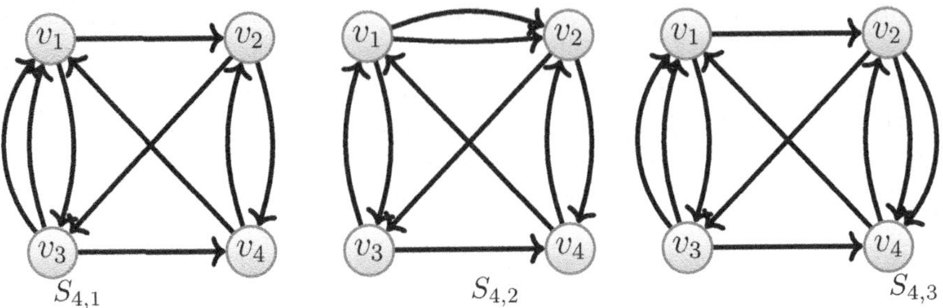

Fig. 5.2 The digraphs $S_{4,1}$, $S_{4,2}$, $S_{4,3}$

of Theorem 5.7, only one of H_1, H_2 and H_3 has an arc. Without loss of generality, assume that H_1 has an arc. Suppose that H_1 has two arcs. Then $H_1 = \overrightarrow{C}_2$. Then we can use the arcs of H_1 to make D_2 strong instead of arcs e_p and e_q used in Case 2 of Part (b) of the proof of Theorem 5.7. Thus, Q has a strong arc decomposition, a contradiction. Hence, if H_1 has an arc, it must have just one arc. This concludes our proof. □

Bang-Jensen, Gutin and Yeo solved an open problem in [75] by obtaining a characterization of all semicomplete compositions with a strong arc decomposition (Theorem 5.13).

Among their argument, the main technical result is the following Theorem 5.11. In order to prove Theorem 5.11, we need the following results.

Lemma 5.13 ([15]) *Let $Q = T[H_1, \ldots, H_t]$ be 2-arc-strong ($t \geq 2$), where T is a semicomplete digraph and every H_i is an arbitrary digraph. If Q contains a cut-vertex, then Q has a strong arc decomposition.*

Lemma 5.14 ([15]) *Let $Q = T[H_1, \ldots, H_t]$ be 2-arc-strong extended semicomplete digraph which has no cut-vertex, $V(T) = \{u_i \mid i \in [t]\}$, $|V(H_i)| \leq 2$ for $i \in [t]$ and $V(H_r) = \{x, y\}$. Suppose $Q' = Q - y$ is not 2-arc-strong. Then there exists an index $q \neq r$ such that one of the following holds:*

(a) $u_r u_q$ is a cut-arc of T and $N^-(V(H_q)) = V(H_r)$,
(b) $u_q u_r$ is a cut-arc of T and $N^+(V(H_q)) = V(H_r)$.

Theorem 5.9 ([15]) *A 2-arc-strong semicomplete directed multigraph D has a strong arc decomposition if and only if it is not isomorphic to one of the exceptional digraphs depicted in Figs. 5.1 and 5.2. Furthermore, a strong arc decomposition of D can be obtained in polynomial time when it exists.*

A *vertex decomposition* of a digraph D is a partition $(V_1, \ldots, V_p)$, $p \geq 1$, of its vertex set. The *index* of vertex v in the decomposition, denoted by $ind(v)$, is the integer i such that $v \in V_i$. An arc uv is *forward* if $ind(u) < ind(v)$, *backward* if $ind(u) > ind(v)$. A vertex decomposition $(V_1, \ldots, V_p)$ is *strong* if $D[V_i]$ is strong for all $i \in [p]$. A *nice vertex decomposition* of a digraph D is a strong decomposition such that the set of cut-arcs of D is exactly the set of backward arcs.

Theorem 5.10 ([16]) *Every strong semicomplete digraph D of order at least 4 admits a nice vertex decomposition $(V_1, \ldots, V_p)$ and this decomposition is unique. The sets $V_i, i \in [p]$, are exactly the strong components of the digraph H that we obtain from D by deleting all cut-arcs.*

By Lemmas 5.13 and 5.14 and Theorems 5.9 and 5.10, Bang-Jensen, Gutin and Yeo obtained Theorem 5.11. Recall that S_4 is shown in Fig. 5.1.

Theorem 5.11 ([15]) *Let $Q = T[\overline{K}_{n_1}, \ldots, \overline{K}_{n_t}]$ be a 2-arc-strong extended semicomplete digraph where $n_i \leq 2$ for $i \in [t]$. If Q is not isomorphic to one of the following three digraphs: S_4, $\overrightarrow{C}_3[\overline{K}_2, \overline{K}_2, \overline{K}_2]$, $\overrightarrow{C}_3[\overline{K}_2, \overline{K}_2, \overrightarrow{P}_2]$, then Q has a strong arc decomposition.*

Theorem 5.12 can be proved by Theorem 5.11.

Theorem 5.12 ([15]) *Let $Q = T[\overline{K}_{n_1}, \ldots, \overline{K}_{n_t}]$ be a 2-arc-strong extended semicomplete digraph. If Q is not isomorphic to one of the following four digraphs: S_4, $\overrightarrow{C}_3[\overline{K}_2, \overline{K}_2, \overline{K}_2]$, $\overrightarrow{C}_3[\overrightarrow{P}_2, \overline{K}_2, \overline{K}_2]$, $\overrightarrow{C}_3[\overline{K}_2, \overline{K}_2, \overline{K}_3]$, then Q has a strong arc decomposition.*

Lemma 5.15 ([15]) *Let $Q = T[H_1, \ldots, H_t]$ be 2-arc-strong $(t \geq 2)$, where T is a semicomplete digraph and every H_i is an arbitrary digraph. Then at least one of the following cases holds.*

(a) Q has a strong arc decomposition.
(b) Q is an extended semicomplete digraph.
(c) For every $i \in [t]$ and every arc e of H_i, $Q - e$ is 2-arc-strong.

Theorem 5.12 and Lemma 5.15 imply Theorem 5.13.

Theorem 5.13 ([15]) *Let $Q = T[H_1, \ldots, H_t]$ be a semicomplete composition. Then Q has a strong arc decomposition if and only if Q is 2-arc-strong and is not isomorphic to one of the following four digraphs: S_4, $\overrightarrow{C}_3[\overline{K}_2, \overline{K}_2, \overline{K}_2]$, $\overrightarrow{C}_3[\overrightarrow{P}_2, \overline{K}_2, \overline{K}_2]$, $\overrightarrow{C}_3[\overline{K}_2, \overline{K}_2, \overline{K}_3]$.*

The following theorem by Bang-Jensen and Huang gives a complete characterization of quasi-transitive digraphs and the decomposition below is called the *canonical decomposition* of a quasi-transitive digraph.

Theorem 5.14 ([8]) *Let D be a quasi-transitive digraph. Then the following assertions hold:*

(a) If D is not strong, then there exists a transitive oriented graph T with vertices $\{u_i \mid i \in [t]\}$ and strong quasi-transitive digraphs $H_1, H_2, \ldots, H_t$ such that $D = T[H_1, H_2, \ldots, H_t]$, where H_i is substituted for u_i, $i \in [t]$.
(b) If D is strong, then there exists a strong semicomplete digraph S with vertices $\{v_j \mid j \in [s]\}$ and quasi-transitive digraphs $Q_1, Q_2, \ldots, Q_s$ such that Q_j is

either a vertex or is non-strong and $D = S[Q_1, Q_2, \ldots, Q_s]$*, where* Q_j *is substituted for* v_j*,* $j \in [s]$*.*

Theorems 5.13 and 5.14 imply a characterization of quasi-transitive digraphs with a strong arc decomposition (this solves another open question in [75]).

Theorem 5.15 ([15]) *A quasi-transitive digraph D has a strong arc decomposition if and only if D is 2-arc-strong and is not isomorphic to one of the following four digraphs:* S_4, $\vec{C}_3[\overline{K}_2, \overline{K}_2, \overline{K}_2]$, $\vec{C}_3[\vec{P}_2, \overline{K}_2, \overline{K}_2]$, $\vec{C}_3[\overline{K}_2, \overline{K}_2, \overline{K}_3]$.

All proofs in [15] are constructive and can be turned into polynomial algorithms for finding strong arc decompositions. Thus, the problem of finding a strong arc decomposition in a semicomplete composition, which has one, admits a polynomial time algorithm.

By the definitions and Theorem 5.14, strong semicomplete compositions generalize both strong semicomplete digraphs and strong quasi-transitive digraphs. However, they do not generalize locally semicomplete digraphs and their generalizations, locally in- and out-semicomplete digraphs. While there is a characterization of locally semicomplete digraphs having a strong arc decomposition (Theorem 5.6), no such a characterization is known for locally in-semicomplete digraphs and it would be interesting to obtain such a characterization or at least establish the complexity of deciding whether an locally in-semicomplete digraph has a strong arc decomposition.

Problem 5.1 ([15]) Can we decide in polynomial time whether a given locally in-semicomplete digraph has a strong arc decomposition?

Problem 5.2 ([15]) Characterize locally in-semicomplete digraphs with a strong arc decomposition.

Recall that a digraph D has a strong arc decomposition if and only if $\lambda_{V(D)}(D) \geq 2$ (or, $\kappa_{V(D)}(D) \geq 2$). Therefore, the problem of ASSP (or ISSP) could be seen as an extension of the problem of strong arc decomposition. Sun, Gutin and Zhang gave two sufficient conditions for a digraph composition to have at least n_0 arc-disjoint S-strong subgraphs for any $S \subseteq V(Q)$ with $2 \leq |S| \leq |V(Q)|$ (Theorem 5.18), where $n_0 = \min\{n_i \mid i \in [t]\}$. To prove Theorem 5.18, we still need the following several results.

Ng [57] proved the following result on the Hamiltonian decomposition of complete regular multipartite digraphs.

Theorem 5.16 ([57]) *The digraph* $\overleftrightarrow{K}_{r,r,\ldots,r}$ *(s times) is Hamiltonian decomposable if and only if* $(r, s) \neq (4, 1)$ *and* $(r, s) \neq (6, 1)$*.*

By Theorem 5.16, we can determine the precise value for the strong subgraph k-arc-connectivity (the definition can be found in Sect. 4.4) of a complete bipartite digraph.

Lemma 5.16 ([77]) *For two positive integers a and b with* $a \leq b$*, we have*

$$\lambda'_k(\overleftrightarrow{K}_{a,b}) = a$$

for $2 \le k \le a+b$.

Proof Let $V(\overleftrightarrow{K}_{a,b}) = V_1 \cup V_2$ with $V_1 = \{u_i \mid i \in [a]\}$ and $V_2 = \{v_j \mid j \in [b]\}$. By Theorem 5.16, the subgraph of $\overleftrightarrow{K}_{a,b}$ induced by $\{u_i, v_j \mid i, j \in [a]\}$ can be decomposed into a Hamiltonian cycles: H_i $(i \in [a])$. For each $i \in [a]$, let D_i be the strong spanning subgraph of $\overleftrightarrow{K}_{a,b}$ obtained from H_i by adding the arc set $\{u_iv_j, v_ju_i \mid a+1 \le j \le b\}$. Observe that these subgraphs are pairwise arc-disjoint, and so $\lambda'_{a+b}(\overleftrightarrow{K}_{a,b}) \ge a$. It is known [68] that $\lambda'_{k+1}(D) \le \lambda'_k(D)$ $(k \in [n-1])$ and $\lambda'_k(D) \le \min\{\delta^+(D), \delta^-(D)\}$ for a digraph D with order n, we have that $a = \min\{\delta^+(\overleftrightarrow{K}_{a,b}), \delta^-(\overleftrightarrow{K}_{a,b})\} \ge \lambda'_2(\overleftrightarrow{K}_{a,b}) \ge \dots \ge \lambda'_{a+b}(\overleftrightarrow{K}_{a,b}) \ge a$. This completes the proof. □

The following result on Hamiltonian decomposition was also proved by Ng.

Lemma 5.17 ([58]) *For any two integers* $t \ge 2$ *and* $r \ge 3$*, the product digraph* $\overrightarrow{C}_t \circ \overline{K_r}$ *is Hamiltonian decomposable. Furthermore, the Hamiltonian cycles in such a decomposition can be found in* $O(n^2)$ *time.*

Recall that a strong semicomplete digraph is also locally in-semicomplete. By Camion's Theorem (see [18] or Theorem 2.2.6 of [6]), there is a Hamiltonian cycle in a strong semicomplete digraph. In fact, Bang-Jensen and Hell obtained a stronger result.

Theorem 5.17 ([7]) *There is an* $O(m + n\log n)$ *algorithm for finding a Hamiltonian cycle in a strong locally in-semicomplete digraph.*

By Lemma 5.17 and Theorem 5.17, the following result holds:

Lemma 5.18 ([77]) *Let* $Q = D[H_1, \dots, H_t]$ *with* $|D| = t \ge 2$ *and* $|V(H_i)| \ge 3$ *for each* $i \in [t]$*. If* D *is a strong semicomplete digraph, then* Q *has at least* n_0 *arc-disjoint strong spanning subgraphs. Moreover, these strong subgraphs can be found in time* $O(n^2)$*, where* n *is the order of* Q*.*

Proof By Theorem 5.17, we can find a Hamiltonian cycle of D in time $O(n^2)$. Clearly, Q contains $\overrightarrow{C}_t \circ \overline{K_{n_0}}$ as a spanning subgraph, where $t \ge 2$. Without loss of generality, assume that $n_0 = |V(H_1)|$. By Lemma 5.17, $\overrightarrow{C}_t \circ \overline{K_{|V(H_1)|}}$ is Hamiltonian decomposible, and these Hamiltonian cycles can be found in time $O(n^2)$. Furthermore, these cycles are desired strong spanning subgraphs in Q. □

Now we are ready to prove Theorem 5.18:

Theorem 5.18 ([77]) *The digraph composition* $Q = D[H_1, \dots, H_t]$ *has at least* n_0 *arc-disjoint* S*-strong subgraphs for any* $S \subseteq V(Q)$ *with* $2 \le |S| \le |V(Q)|$*, if one of the following conditions holds:*

(a) D is a strong symmetric digraph;
(b) D is a strong semicomplete digraph and $Q \notin \mathcal{Q}_0$, where
$\mathcal{Q}_0 = \{\overrightarrow{C}_3[\overline{K}_2, \overline{K}_2, \overline{K}_2], \overrightarrow{C}_3[\overrightarrow{P}_2, \overline{K}_2, \overline{K}_2], \overrightarrow{C}_3[\overline{K}_2, \overline{K}_2, \overline{K}_3]\}$.

Moreover, these strong subgraphs can be found within time complexity $O(n^4)$, where n is the order of Q.

Proof **Part (a)** For any $S \subseteq V(Q)$ with $2 \leq |S| \leq |V(Q)|$, we will obtain n_0 arc-disjoint S-strong subgraphs using the following three steps:
Step 1: We obtain a spanning subgraph $Q' = D[H'_1, \dots, H'_t]$ of Q such that $V(H'_i) = V(H_i)$ and each H'_i has no arcs, where $i \in [t]$.
Step 2: For each pair of $p, q \in [t]$ such that $u_pu_q, u_qu_p \in A(D)$, let $Q_{p,q}$ be the subgraph of Q' induced by the vertex set $\{u_{p,j_p}, u_{q,j_q} \mid j_p \in [n_p], j_q \in [n_q]\}$. We obtain n_0 arc-disjoint spanning strong subgraphs: $\{D_{p,q,s} \mid s \in [n_0]\}$ in $Q_{p,q}$ by the construction of Lemma 5.16, since each $Q_{p,q}$ is a complete bipartite digraph.
Step 3: For each $s \in [n_0]$, let D_s be the union of all $D_{p,q,s}$ with $u_pu_q, u_qu_p \in A(D)$.

Observe that the subgraphs in Step 3 are strong and pairwise arc-disjoint, so we obtain a set of n_0 arc-disjoint spanning strong subgraphs of Q', furthermore, these subgraphs are desired arc-disjoint S-strong subgraphs of Q.

Step 1 can be performed in $O(n^2)$ time. In Step 2, there are at most $\binom{n}{2}$ pairs of p, q, and note that $\{D_{p,q,s} \mid s \in [n_0]\}$ can be found in $O(n^2)$ time in $Q_{p,q}$ by the construction of Lemma 5.16, so Step 2 can be executed in time $O(n^4)$. Step 3 can be performed in time $O(n^2)$. Hence the desired subgraphs can be found in polynomial time $O(n^4)$. This completes the proof of part (a).

Part (b) For the case that $n_0 = 1$, Q itself is the desired strong subgraph. The result holds for the case that $n_0 = 2$ by Theorem 5.8 and the fact that a strong spanning subgraph is an S-strong subgraph for any $S \subseteq V(Q)$ with $2 \leq |S| \leq |V(Q)|$. It follows from the construction proof of Theorem 5.8 that these spanning strong subgraphs can be found in $O(n^3)$ time.

For the case that $n_0 \geq 3$, we will get n_0 arc-disjoint S-strong subgraphs for any $S \subseteq V(Q)$ with $2 \leq |S| \leq |V(Q)|$ by the following two steps:
Step 1: Find n_0 arc-disjoint spanning strong subgraphs: $D'_1, \dots, D'_{n_0}$ in Q' by Lemma 5.18, where $Q' = D[H'_1, \dots, H'_t]$ is an induced subgraph of Q such that $V(H'_i) = \{u_{i,j_i} \mid i \in [t], j_i \in [n_0]\}$.
Step 2: For each $j \in [n_0]$, we construct a spanning subgraph D_j of Q from D'_j by adding arcs between $V(Q) \setminus V(Q')$ and $\{u_{i,j} \mid i \in [t]\}$.

Observe that these subgraphs in Step 2 are strong and pairwise arc-disjoint, so we obtain a set of n_0 arc-disjoint spanning strong subgraphs of Q, furthermore, these subgraphs are desired arc-disjoint S-strong subgraphs. Step 1 can be performed in $O(n^2)$ time by Lemma 5.17 and Step 2 can be performed in $O(n^3)$ time. This completes the proof of part (b). □

By Theorems 5.14 and 5.18, we directly have:

Corollary 5.4 ([77]) *Let $Q \notin \mathcal{Q}_0$ be a strong quasi-transitive digraph. We can in polynomial time find at least n_0 arc-disjoint S-strong subgraphs in Q for any $S \subseteq V(Q)$ with $2 \leq |S| \leq |V(Q)|$.*

5.4 Digraph Products

In the arguments of this section, we will use the following terminology and notation. Let G and H be two digraphs with $V(G) = \{u_i \mid i \in [n]\}$ and $V(H) = \{v_j \mid j \in [m]\}$. For simplicity, we let $u_{i,j} = (u_i, v_j)$ for $i \in [n], j \in [m]$. We use $G(v_j)$ to denote the subdigraph of $G\Box H$ induced by vertex set $\{u_{i,j} \mid i \in [n]\}$, where $j \in [m]$, and use $H(u_i)$ to denote the subdigraph of $G\Box H$ induced by vertex set $\{u_{i,j} \mid j \in [m]\}$, where $i \in [n]$. Clearly, we have $G(v_j) \cong G$ and $H(u_i) \cong H$. (For example, as shown in Fig. 5.3, $G(v_j) \cong G$ for $j \in [4]$ and $H(u_i) \cong H$ for $i \in [3]$.) For $1 \leq j_1 \neq j_2 \leq m$, u_{i,j_1} and u_{i,j_2} belong to the same digraph $H(u_i)$, where $u_i \in V(G)$; we call u_{i,j_2} the *vertex corresponding to* u_{i,j_1} in $G(v_{j_2})$; for $1 \leq i_1 \neq i_2 \leq n$, we call $u_{i_2,j}$ the vertex corresponding to $u_{i_1,j}$ in $H(u_{i_2})$. Similarly, we can define the subdigraph *corresponding* to some other subdigraph. For example, in Fig. 5.3c, let P_1 (P_2) be the path labelled 1 (2) in $H(u_1)$ ($H(u_2)$), then P_2 is called the path *corresponding* to P_1 in $H(u_2)$. The readers can see [39] for more information on digraph products.

Lemma 5.19 ([75]) *For any integer $n \geq 2$, the product digraph $D = \overrightarrow{C}_n\Box\overrightarrow{C}_n$ can be decomposed into two arc-disjoint Hamiltonian cycles.*

Proof Let $G = H \cong \overrightarrow{C}_n$; moreover $G = u_1u_2\dots u_nu_1$ and $H = v_1v_2\dots v_nv_1$. Let $P_i = G(v_i) - u_{n-i,i}u_{n+1-i,i}$ for $i \in [n-1]$ and $P_n = G(v_n) - u_{n,n}u_{1,n}$. Let $Q_i = H(u_i) - u_{i,n-i}u_{i,n+1-i}$ for $i \in [n-1]$ and $Q_n = H(u_n) - u_{n,n}u_{n,1}$. Furthermore, let

$$D' = \left(\bigcup_{i=1}^{n-1}(P_i \cup \{u_{n-i,i}u_{n-i,i+1}\})\right) \cup (P_n \cup \{u_{n,n}u_{n,1}\})$$

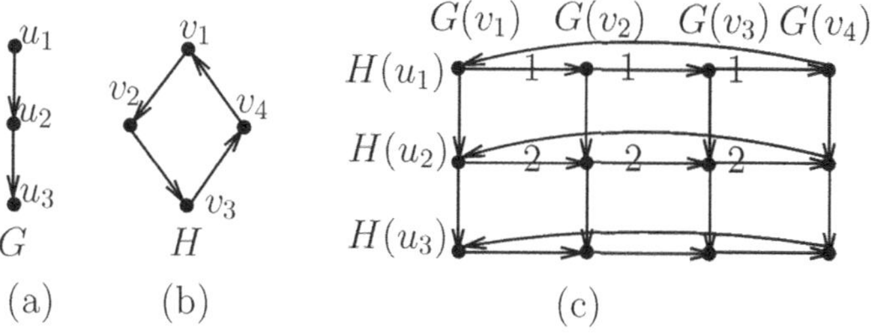

Fig. 5.3 Two digraphs G, H and their Cartesian product. (**a**) A directed path G with length two (**b**) a directed cycle H with length four (**c**) the Cartesian product of G and H

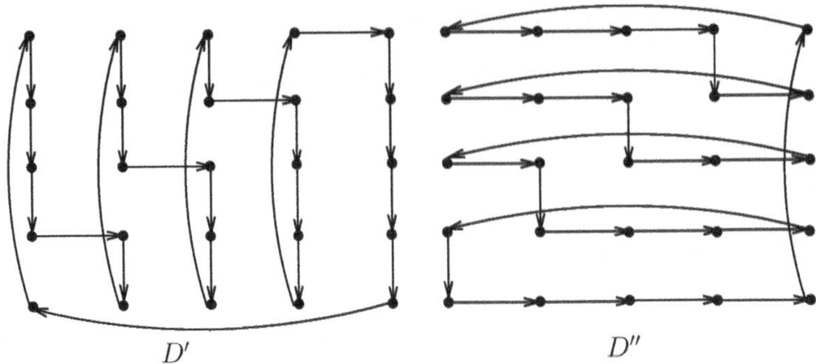

Fig. 5.4 Two arc-disjoint Hamiltonian cycles for the case $n = 5$

and

$$D'' = \left(\bigcup_{i=1}^{n-1}(Q_i \cup \{u_{i,n-i}u_{i+1,n-i}\})\right) \cup (Q_n \cup \{u_{n,n}u_{1,n}\})$$

By the construction, the subdigraphs D' and D'' are Hamiltonian cycles of D. For example, see Fig. 5.4 for the case that $n = 5$ (the Hamiltonian cycle D' consists of five "vertical" paths P_i of order five and five "horizontal" arcs, D'' consists of five "horizontal" paths Q_i of order five and five "vertical" arcs, furthermore, these two cycles are symmetric about the diagonal.) □

Note that deciding whether a digraph D has a collection of arc-disjoint cycles covering all vertices of D can be done in polynomial time using network flows. Indeed, assign lower bound 1 and upper bound $\min\{d^-(x), d^+(x)\}$ to every vertex x in D and lower bound 0 and upper bound 1 to every arc of D. Observe that the resulting network has a feasible flow if and only if D has a collection of arc-disjoint cycles covering all vertices of D. Observe that the existence of a flow in a network with lower and upper bounds on vertices and arcs can be decided in polynomial time, see e.g., Chapter 4 in [5]. Moreover, we can compute such a flow in polynomial time (if it exists) and obtain the corresponding collection of cycles in D. The following lemma may be of independent interest.

Lemma 5.20 ([75]) *Let G be a strong digraph with at least two vertices which has a collection of arc-disjoint cycles covering all its vertices. Then the product digraph $D = G\Box G$ has a strong arc decomposition. Moreover, such a strong arc decomposition can be found in polynomial time.*

Proof By the arguments in the paragraph before this lemma, we may assume that we are given a collection $(P_0, P_1, P_2, \ldots, P_p)$ of arc-disjoint cycles covering all vertices of G. For each $h \in \{0, 1, 2, \ldots, p\}$, let G_h denote the digraph with vertices

$\bigcup_{i=0}^{h} V(P_i)$ and arcs $\bigcup_{i=0}^{h} A(P_i)$. Now we will prove the lemma by induction on the number of cycles in the collection.

For the base step, by Lemma 5.19, we have that $G_0 \Box G_0 = P_0 \Box P_0$ can be decomposed into two arc-disjoint strong spanning subdigraphs.

For the inductive step, we assume that $G_h \Box G_h$ $(0 \le h \le p-1)$ can be decomposed into two arc-disjoint strong spanning subdigraphs D'_h and D''_h. We will construct two arc-disjoint spanning strong subgraphs in $G_{h+1} \Box G_{h+1}$.

If $V(G_h) \subseteq V(P_{h+1})$, then P_{h+1} is a Hamiltonian cycle of G_{h+1}, and we are done by Lemma 5.19. If $V(P_{h+1}) \subseteq V(G_h)$, then G_h is a strong spanning subdigraph of G_{h+1}, and we are also done by the induction hypothesis.

In the following argument, we assume that $V(G_h) \setminus V(P_{h+1}) \neq \emptyset$ and $V(P_{h+1}) \setminus V(G_h) \neq \emptyset$. Without loss of generality, for the first copies of G_h and P_{h+1} in $G_h \Box G_h$ and $P_{h+1} \Box P_{h+1}$, let $V(G_h) = \{u_i \mid i \in [t]\}$, $V(P_{h+1}) = \{u_i \mid s \le i \le \ell\}$. We have $1 < s \le t < \ell$. For the second copies of G_h and P_{h+1} in $G_h \Box G_h$ and $P_{h+1} \Box P_{h+1}$, we will use v_i's rather than u_i's.

By Lemma 5.19, in $G_{h+1} \Box G_{h+1}$, the subdigraph $P_{h+1} \Box P_{h+1}$ can be decomposed into two arc-disjoint spanning strong subgraphs $\overline{D}'_h$ and $\overline{D}''_h$. Observe that

$$V(G_h \Box G_h) \cap V(P_{h+1} \Box P_{h+1}) \supseteq \{u_{t,t}\} \text{ and } A(G_h \Box G_h) \cap A(P_{h+1} \Box P_{h+1}) = \emptyset.$$

For $j \in [s-1]$, let $G_{h,j}$ be the subdigraph of $G(v_j)$ corresponding to P_{h+1}. For $t+1 \le j \le \ell$, let $G_{h,j}$ be the subdigraph of $G(v_j)$ corresponding to G_h. For $i \in [s-1]$, let $H_{h,i}$ be the subdigraph of $H(u_i)$ corresponding to P_{h+1}. For $t+1 \le i \le \ell$, let $H_{h,i}$ be the subdigraph of $H(u_i)$ corresponding to G_h.

Now let D'_{h+1} be a union of the following strong digraphs: D'_h, $\overline{D}'_h$, $H_{h,i}$ and $G_{h,j}$ for all $t+1 \le i, j \le \ell$. Observe that D'_{h+1} is a strong spanning subdigraph of $G_{h+1} \Box G_{h+1}$ since $\overline{D}'_h$ has at least one common vertex with each of D'_h, $H_{h,i}$ and $G_{h,j}$ for all $t+1 \le i, j \le \ell$. Let D''_{h+1} be a spanning subdigraph of $G_{h+1} \Box G_{h+1}$ with $A(D''_{h+1}) = A(G_{h+1} \Box G_{h+1}) \setminus A(D'_{h+1})$. Observe that D''_{h+1} is the union of D''_h, $\overline{D}''_h$, $H_{h,i}$ and $G_{h,j}$ for all $1 \le i, j \le s-1$. And D''_h has at least one common vertex with each of $\overline{D}''_h$, $H_{h,i}$ and $G_{h,j}$ for all $1 \le i, j \le s-1$, thus D''_{h+1} is strong.

Hence, we complete the inductive step and conclude that $D = G \Box G$ can be decomposed into two arc-disjoint strong spanning subdigraphs. Moreover, by the above argument, these subdigraphs can be found in polynomial time. □

Lemma 5.21 ([75]) *For any two strong digraphs G and H, if G has a strong arc decomposition, then the product digraph $D = G \Box H$ has a strong arc decomposition.*

Proof Let $V(G) = \{u_i \mid i \in [n]\}$, $V(H) = \{v_j \mid j \in [m]\}$, and G contain two arc-disjoint spanning strong subgraphs G_1 and G_2. For $i \in [2]$ and $j \in [m]$, let $G_{i,j}$ be the subdigraph of $G(v_j)$ corresponding to G_i. As shown in Fig. 5.5, let D' be the union of $H(u_1)$ and $G_{1,j}$ for all $j \in [m]$, and let D'' be a subdigraph of D with $V(D'') = V(D)$ and $A(D'') = A(D) \setminus A(D')$. Note that D'' is the union of $H(u_i)$

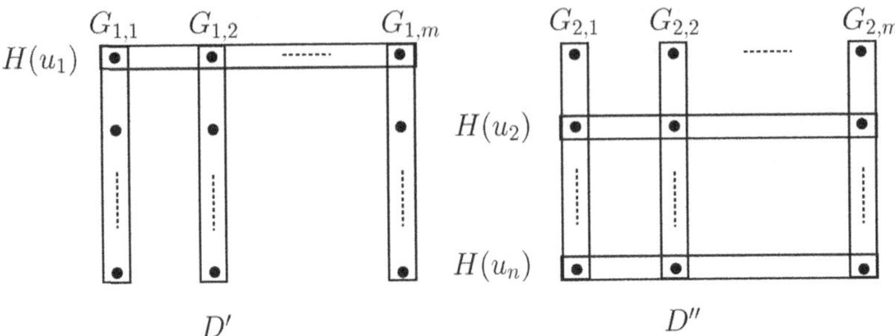

Fig. 5.5 Two arc-disjoint spanning strong subgraphs in Lemma 5.21

and $G_{2,j}$ for all $2 \le i \le n$, $j \in [m]$. Since $H(u_i)$, $G_{1,j}$ and $G_{2,j}$ $(i \in [n], j \in [m])$ are strong, both D' and D'' are strong spanning subdigraphs of D. This completes the proof. □

By the definition of the kth power with respect to Cartesian product $D^{\square k}$, the associativity of the Cartesian product (up to isomorphism), and Lemmas 5.20 and 5.21, we can obtain the following result on $G^{\square k}$ for any integer $k \ge 2$.

Theorem 5.19 ([75]) *Let G be a strong digraph of order at least two which has a collection of arc-disjoint cycles covering all its vertices and let $k \ge 2$ be an integer. Then the product digraph $D = G^{\square k}$ has a strong arc decomposition. Moreover, for any fixed integer k, such a strong arc decomposition can be found in polynomial time.*

By definition, $G\square H$ is a spanning subdigraph of $G \boxtimes H$, and $G \boxtimes H$ is strongly connected if and only if both G and H are strongly connected [39]. In the following argument, we will still use the terminology and notation introduced earlier in this section, since $G\square H$ is a spanning subdigraph of $G \boxtimes H$.

The following result on strong product of two cycles was obtained by Sun, Gutin and Ai.

Lemma 5.22 ([75]) *For any two integers $n, m \ge 2$, the product digraph $D = \overrightarrow{C}_n \boxtimes \overrightarrow{C}_m$ has a strong arc decomposition.*

Proof Let $\overrightarrow{C}_n = u_1u_2 \dots u_nu_1$ and $\overrightarrow{C}_m = v_1v_2 \dots v_mv_1$. Let D' be the spanning subdigraph of D which is the union of $G(v_j)$ for $j \in [m]$ and the following additional m arcs: $\{u_{n,j}u_{1,j+1} \mid j \in [m-1]\} \cup \{u_{1,m}u_{2,1}\}$. Observe that D' is strong. Let D'' be a spanning subdigraph of D with $A(D'') = A(D) \setminus A(D')$. To see that D'' is strong, observe that it contains $H(u_i)$ for $i \in [n]$ and arcs $\{u_{i,1}u_{i+1,2} \mid i \in [n-1]\} \cup \{u_{n,m}u_{1,1}\}$. □

The following result on ear decomposition is well-known, see e.g., [5].

Theorem 5.20 *Let D be a digraph with at least two vertices. Then D is strong if and only if it has an ear decomposition. Furthermore, if D is strong, every cycle can be used as a starting cycle P_0 for an ear decomposition of D, and there is a linear-time algorithm to find such an ear decomposition.*

By Lemma 5.22 and Theorem 5.20, Sun, Gutin and Ai obtained the following result on strong arc decomposition of strong product digraphs.

Theorem 5.21 ([75]) *For any strong digraphs G and H with at least two vertices, the product digraph $D = G \boxtimes H$ has a strong arc decomposition. Moreover, such a decomposition can be found in polynomial time.*

Proof By Theorem 5.20, G has an ear decomposition $\mathcal{P} = (P_0, P_1, P_2, \ldots, P_p)$ and H has an ear decomposition $\mathcal{Q} = (Q_0, Q_1, Q_2, \ldots, Q_q)$, such that P_0 is a cycle of G and Q_0 is a cycle of H. Let G_i denote the subdigraph of G with vertices $\bigcup_{j=0}^{i} V(P_j)$ and arcs $\bigcup_{j=0}^{i} A(P_j)$ and let H_i denote the subdigraph of H with vertices $\bigcup_{j=0}^{i} V(Q_j)$ and arcs $\bigcup_{j=0}^{i} A(Q_j)$.

We will prove the theorem by induction on $r \in \{0, 1, \ldots, p+q\}$. For the base step, by Lemma 5.22, we have that $P_0 \boxtimes Q_0$ has a strong arc decomposition. For the inductive step, we assume that $r = h + g < p + q$ ($h \leq p, g \leq q$) and $G_h \boxtimes H_g$ has a strong arc decomposition (D', D'').

Since strong product is a commutative operation, without loss of generality it suffices to prove that $G_{h+1} \boxtimes H_g$ ($h < p$) has a strong arc decomposition. Let $V(G_h) = \{u_1, u_2, \ldots, u_\ell\}$, $V(H_g) = \{v_1, v_2, \ldots, v_m\}$ and $v_1v_s \in A(H_g)$. Let $P_{h+1,j}$ be the subdigraph of $G(v_j)$ corresponding to P_{h+1} for $j \in [m]$. We will consider two cases.

Case 1 P_{h+1} is a cycle.

Let $P_{h+1} = u_\ell u_{\ell+1} \ldots u_n u_\ell$. Observe that every $P_{h+1,j}$ for $j \in [m]$ shares vertex $u_{\ell,j}$ with D'. Thus, the union U_1 of D' and $P_{h+1,j}$ for $j \in [m]$ is a spanning strong subgraph of $G_{h+1} \boxtimes H_g$. Let $V(U_2) = V(G_{h+1} \boxtimes H_g)$ and $A(U_2) = A(G_{h+1} \boxtimes H_g) \setminus A(U_1)$.

Observe that $A(U_2)$ contains $A(D'')$, $A(H(u_i))$ for $\ell + 1 \leq i \leq n$ and $\{u_{i,1}u_{i+1,s} \mid \ell \leq i \leq n-1\} \cup \{u_{n,1}u_{\ell,s}\}$. Thus, U_2 is strong.

Case 2 P_{h+1} is a path.

Let $P_{h+1} = u_\ell u_{\ell+1} \ldots u_{n-1}u_t$, where $t < \ell$. Let U_1 be the union of D' and $P_{h+1,j}$ for $j \in [m]$. Observe that U_1 is a spanning subdigraph of $G_{h+1} \boxtimes H_g$ and strong since every $P_{h+1,j}$ for $j \in [m]$ shares its end-vertices with D'. Let $V(U_2) = V(G_{h+1} \boxtimes H_g)$ and $A(U_2) = A(G_{h+1} \boxtimes H_g) \setminus A(U_1)$. Observe that $A(U_2)$ contains $A(D'')$, $A(H(u_i))$ for $\ell + 1 \leq i \leq n-1$ and $\{u_{i,1}u_{i+1,s} \mid \ell \leq i \leq n-2\} \cup \{u_{n-1,1}u_{t,s}\}$. Thus, U_2 is strong.

Hence, we complete the inductive step and conclude that $D = G \boxtimes H$ has a strong arc decomposition. Furthermore, by Theorem 5.20, the proof of Lemma 5.22, and the argument of this theorem, we can conclude that such a decomposition can be found in polynomial time. □

By definition, the strong product digraph $G \boxtimes H$ is a spanning subdigraph of the lexicographic product digraph $G \circ H$, so the following result holds by Theorem 5.21: For any strong connected digraphs G and H with orders at least 2, the product digraph $D = G \circ H$ has a strong arc decomposition. Moreover, such a decomposition can be found in polynomial time. In fact, we can get a more general result.

For the Hamiltonian decomposition problem, Ng [58] gave the most complete result among digraph products.

Theorem 5.22 ([58]) *If G and H are Hamiltonian decomposable digraphs, and $|V(G)|$ is odd, then $G \circ H$ is Hamiltonian decomposable.*

Theorem 5.22 implies that if G and H are Hamiltonian decomposable digraphs, and $|V(G)|$ is odd, then $G \circ H$ has a strong arc decomposition. It is not hard to extend this result as follows: for any strong digraphs G and H of orders at least 2, if H contains $\ell \geq 1$ arc-disjoint spanning strong subgraphs, then the product digraph $D = G \circ H$ can be decomposed into $\ell + 1$ arc-disjoint spanning strong subgraphs.

Trotter and Erdős gave a characterization for the Hamiltonicity of the Cartesian product of two cycles.

Theorem 5.23 ([80]) *The Cartesian product $\overrightarrow{C}_p \Box \overrightarrow{C}_q$ is Hamiltonian if and only if there are non-negative integers d_1, d_2 for which $d_1 + d_2 = \gcd(p, q) \geq 2$ and $\gcd(p, d_1) = \gcd(q, d_2) = 1$.*

It was proved in [75] that $G \Box H$ has a strong arc decomposition when $G \cong H$ and G contains a collection of arc-disjoint cycles covering all its vertices (Lemma 5.20). However, Theorem 5.23 implies this cannot be extended to the case that $G \ncong H$, since it is not hard to show that the Cartesian product digraph of any two cycles has a strong arc decomposition if and only if it has a pair of arc-disjoint Hamiltonian cycles. But this could hold for the case that $G \ncong H$ if we have other conditions, since it was also showed in [75] that $G \Box H$ has a strong arc decomposition when one of G and H has a strong arc decomposition (Lemma 5.21). So the following open question is interesting:

Problem 5.3 ([75]) For any two strong digraphs G and H, neither of which have a strong arc decomposition, under what condition the product digraph $G \Box H$ has a strong arc decomposition?

Furthermore, we may also consider the following more challenging question:

Problem 5.4 ([75]) Under what condition the product digraph $G \Box H$ $(G \boxtimes H)$ has more (than two) arc-disjoint spanning strong subgraphs?

5.5 Related Topics: Kelly Conjecture and Bang-Jensen–Yeo Conjecture

The following is the famous Kelly conjecture (see e.g. Conjecture 13.4.5 of [5]):

Conjecture 5.4 Every regular tournament on $n = 2k + 1$ vertices has a decomposition into k-arc-disjoint Hamiltonian cycles.

This conjecture was proved for large n by Kühn and Osthus [51]. It is easy to see that a k-regular tournament is k-arc-strong. Bang-Jensen and Yeo posed the following conjecture which contains Conjecture 5.4 as the special case when $n = 2k + 1$.

Conjecture 5.5 ([12]) Every k-arc-strong tournament has an arc-decomposition into k arc-disjoint spanning strong subgraphs. Furthermore, for every natural number k, there exists a natural number n_k such that every k-arc-strong semicomplete digraph with at least n_k vertices has an arc-decomposition into k arc-disjoint spanning strong subgraphs.

Conjecture 5.5 is equivalent to the following: for a tournament T, $\kappa_{V(T)}(T) \geq k$ if T is k-arc-strong. Bang-Jensen and Yeo proved three results which support the conjecture, the first one is Theorem 5.2 and the remaining two are as follows:

Theorem 5.24 ([12]) *Every tournament which has a non-trivial cut (both sides containing at least 2 vertices) with precisely k arcs in one direction contains k arc-disjoint spanning strong subgraphs.*

Theorem 5.25 ([12]) *Every k-arc-strong tournament with minimum semi-degree at least $37k$ contains k arc-disjoint spanning strong subgraphs.*

Theorem 5.25 implies that if T is a $74k$-arc-strong tournament with specified not necessarily distinct vertices $u_1, u_2, \dots, u_k, v_1, v_2, \dots, v_k$ then T contains $2k$ arc-disjoint branchings $B^-_{u_1}, B^-_{u_2}, \dots, B^-_{u_k}, B^+_{v_1}, B^+_{v_2}, \dots, B^+_{v_k}$ where $B^-_{u_i}$ is an in-branching rooted at the vertex u_i and $B^+_{v_i}$ is an out-branching rooted at the vertex v_i, $i \in [k]$. This solves a conjecture of Bang-Jensen and Gutin [4].

Conjecture 5.5 does not hold for locally semicomplete digraphs as by Theorem 5.6 the second power of an even cycle cannot be decomposed into two arc-disjoint Hamiltonian cycles. If n is relatively prime to both 2 and 3, then it is easy to see that $\overrightarrow{C}^3_n$ can be decomposed into three arc-disjoint Hamiltonian cycles. In fact, such a decomposition does not exist for any other n.

Proposition 5.1 ([9]) *If n is not relatively prime to 2 or 3, then $\overrightarrow{C}^3_n$ cannot be decomposed into arc-disjoint Hamiltonian cycles.*

Chapter 6
Directed Steiner Cycle Packing Problem

Abstract In the two former sections, we introduce the complexity results for IDSCP on Eulerian digraphs and symmetric digraphs, and the complexity result for ADSCP on general digraphs, when both k and ℓ are fixed. In the third section, we introduce a related topic: directed cycle k-connectivity.

6.1 Results for IDSCP

Using Theorem 2.13, Sun and Jin proved the following NP-completeness of deciding whether $\kappa_S^c(D) \geq \ell$ for Eulerian digraphs (and therefore for general digraphs).

Theorem 6.1 ([72]) *Let $k \geq 2, \ell \geq 1$ be fixed integers. For any Eulerian digraph D and $S \subseteq V(D)$ with $|S| = k$, it is NP-complete to decide whether $\kappa_S^c(D) \geq \ell$.*

In their argument, Sun and Jin proved the NP-hardness of this problem by the NP-hardness of DIRECTED 2-LINKAGE problem in Eulerian digraphs (Theorem 2.13). Their construction of the reduction for the case that $k \geq 2, \ell \geq 2$ is as follows [72]:

Let $[H; s_1, s_2, t_1, t_2]$ be any instance of the latter problem such that H is Eulerian, and (s_1, t_1, s_2, t_2) is a sequence of four terminal vertices of H. Let

$$V(H') = V(H) \cup S \cup \{r_1, r_2\},$$

where $S = \{x_i \mid i \in [k]\}$, and let

$$\begin{aligned} A(H') = A(H) &\cup \{x_{k-1}s_1, t_1x_k, x_ks_2, t_2x_1, s_1r_1, r_1t_2, s_2r_2, r_2t_1\} \\ &\cup \{x_ix_{i+1} \mid i \in [k]\}, \end{aligned}$$

where $x_{k+1} = x_1$. Furthermore, we replicate the arc x_ix_{i+1} $\ell - 1$ additional copies for each $i \in [k-2]$ and replicate the two arcs $x_{k-1}x_k$ and x_kx_1 $\ell - 2$ additional copies (note that when $k = 2$, we just replicate the two arcs x_1x_2 and x_2x_1 $\ell - 2$ additional copies). Finally, to avoid parallel arcs, insert a new vertex $z_{i,i+1}^j$ to each

Y. Sun, *Steiner Type Packing Problems in Digraphs*, SpringerBriefs in Mathematics,
https://doi.org/10.1007/978-981-95-8743-8_6

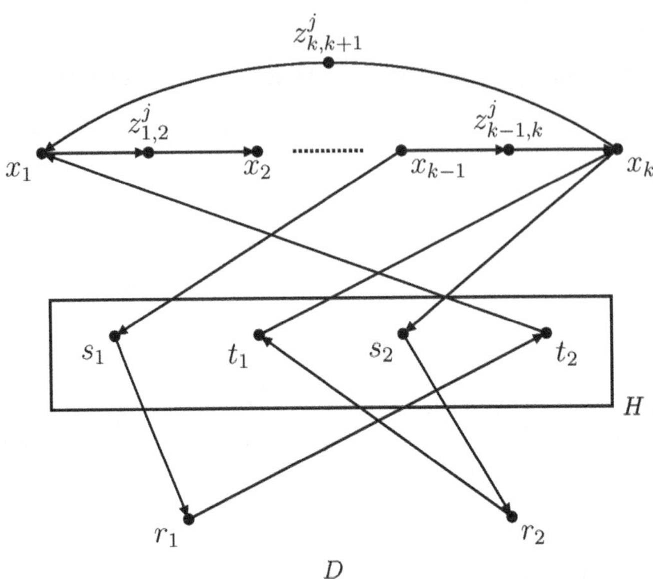

Fig. 6.1 The digraph D

arc $x_i x_{i+1}$, where $j \in [\ell]$ if $i \in [k-2]$ and $j \in [\ell - 1]$ otherwise. The resulting digraph is called D as shown in Fig. 6.1.

When D is an Eulerian digraph, we show in Theorem 6.1 that the problem of deciding whether $\kappa_S^c(D) \geq \ell$ with $|S| = k$ is NP-complete, where both $k \geq 2, \ell \geq 1$ are fixed integers. However, when we consider the class of symmetric digraphs, the problem becomes polynomial-time solvable, as shown in the following theorem which can be deduced from Corollary 2.1.

Theorem 6.2 ([72]) *Let $k \geq 2$ and $\ell \geq 1$ be fixed integers. We can in polynomial time decide if $\kappa_S^c(D) \geq \ell$ for any symmetric digraph D with $S \subseteq V(D)$, where $|S| = k$.*

We directly have the following result by Theorem 6.2.

Corollary 6.1 ([72]) *We can in polynomial time decide if a given symmetric digraph D is k-cyclic, for a fixed integer $k \geq 2$.*

It would also be interesting to study the complexity of IDSCP on other digraph classes, such as semicomplete digraphs.

Problem 6.1 Let $k \geq 2$ and $\ell \geq 1$ be fixed integers. For a semicomplete digraph D and $S \subseteq V(D)$ with $|S| = k$, what is the complexity of deciding whether $\kappa_S^c(D) \geq \ell$?

6.2 Results for ADSCP

Now we turn our attention to the complexity for $\lambda_S^c(D)$. Sun and Jin obtained the following result on general digraphs by Theorem 6.1.

Theorem 6.3 ([72]) *Let $k \geq 2, \ell \geq 1$ be fixed integers. For a digraph D and $S \subseteq V(D)$ with $|S| = k$, it is NP-complete to decide whether $\lambda_S^c(D) \geq \ell$.*

In their argument, Sun and Jin proved the NP-hardness of this problem by reducing from DIRECTED 2-LINKAGE in Eulerian digraphs. Especially, for the case that $k \geq 2$, $\ell \geq 2$, they first produced a digraph D the same as that in Theorem 6.1, and then constructed a new digraph D' from D using a classical reduction (see (70.7) in [62]) as follows: replace every vertex u of H by two vertices u^- and u^+ such that u^-u^+ is an arc in D' and for every $uv \in A(H)$ add an arc u^+v^- to D'. Also, for $z \in S \cup \{r_1, r_2\}$, for every arc zu in D add an arc zu^- to D' and for every arc uz add an arc u^+z to D'.

It would also be interesting to study the complexity of ADSCP on some digraph classes, such as semicomplete digraphs, Eulerian digraphs and symmetric digraphs.

Problem 6.2 Let $k \geq 2$ and $\ell \geq 1$ be fixed integers. For a semicomplete digraph (Eulerian digraph, or symmetric digraph) D and $S \subseteq V(D)$ with $|S| = k$, what is the complexity of deciding whether $\lambda_S^c(D) \geq \ell$?

Kühn and Osthus gave the following sufficient condition for a digraph to be k-cyclic.

Theorem 6.4 ([50]) *Let $k \geq 2$ be an integer. Every digraph D of order $n \geq 200k^3$ which satisfies $\delta^0(D) \geq (n+k)/2 - 1$ is k-cyclic.*

Using Theorem 6.4, we can prove the following:

Proposition 6.1 *Let $k \geq 2, \ell \geq 1$ be integers. Let D be a digraph of order $n \geq 200k^3$ which satisfies $\delta^0(D) \geq (n+k)/2 + \ell - 2$. Then $\lambda_S^c(D) \geq \ell$ for every $S \subseteq V(D)$ with $|S| = k$.*

Proof Let D and S be defined as in the assumption. At stage 1, we set $D_0 = D$. Since $\delta^0(D_0) \geq (n+k)/2 + \ell - 2 \geq (n+k)/2 - 1$, by Theorem 6.4, there is an S-cycle, say C_1, in D_0, then we obtain D_1 from D_0 by deleting the arcs of C_1. Generally, at stage $i \in [\ell]$, we start with a digraph D_{i-1} which is obtained from D_0 by deleting the arcs of $i-1$ arc-disjoint cycles: C_j $(j \in [i-1])$. Since now $\delta^0(D_{i-1}) \geq (n+k)/2+\ell-2-(i-1) = (n+k)/2+\ell-i-1 \geq (n+k)/2-1$, by Theorem 6.4, there is an S-cycle, say C_i, in D_0, then we obtain D_i from D_{i-1} by deleting the arcs of C_i. By induction, we can obtain a set of ℓ arc-disjoint S-cycles: C_j $(j \in [\ell])$. Therefore, $\lambda_S^c(D) \geq \ell$. □

It seems that the bound for $\delta^0(D)$ in Proposition 6.1 is not sharp, so the following question is interesting:

Problem 6.3 Let D be a digraph with order n and minimum semi-degree $\delta^0(D)$. Give a sharp lower bound for $\delta^0(D)$ (in terms of n and k) to guarantee that $\lambda^c_S(D) \geq \ell$ for every $S \subseteq V(D)$ with $|S| = k$.

Let $D = (V(D), A(D))$ be a digraph of order n, $S \subseteq V(D)$ a k-subset of $V(D)$, where $2 \leq k \leq n$. A subset $F \subseteq A(D)$ of arcs is called an *S-feedback arc set* of D if $D \setminus F$ contains no S-cycles. We use $\tau^c_S(D)$ to denote the minimum size of an S-feedback arc set. Clearly, for any $S \subseteq V(D)$, we have $\kappa^c_S(D) \leq \tau^c_S(D)$. It is natural to study the following question:

Problem 6.4 Can we give an upper bound for $\tau^c_S(D)$ by a function of $\kappa^c_S(D)$?

Or, maybe one consider the following Erdős-Pósa type question:

Problem 6.5 Let $D = (V(D), A(D))$ be a digraph of order n, $S \subseteq V(D)$ a k-subset of $V(D)$, where $2 \leq k \leq n$. Is there a function $f(\ell)$ such that the following property holds: If the size of each S-feedback arc set in D is at least $f(\ell)$, then D contains at least ℓ pairwise edge-disjoint S-cycles.

6.3 A Related Topic: Directed Cycle k-Connectivity

By replacing the "S-strong subgraphs" in the definition of strong subgraph k-connectivity with "S-cycles", Wang and Sun [83] obtained the following definition of directed cycle k-connectivity $\kappa^c_k(D)$ of a digraph D. Recall that $\kappa^c_S(D)$ denotes the maximum number of pairwise internally-disjoint S-cycles in D. Let

$$\kappa^c_k(D) = \min\left\{\kappa^c_S(D) \mid S \subseteq V(D), |S| = k, 2 \leq k \leq n\right\}.$$

In [83], Wang and Sun studied the directed cycle k-connectivity of complete digraphs $\overleftrightarrow{K}_n$ and complete regular bipartite digraphs $\overleftrightarrow{K}_{a,a}$. They gave a sharp lower bound for $\kappa^c_k(\overleftrightarrow{K}_n)$ and determine the exact values for $\kappa^c_k(\overleftrightarrow{K}_n)$ when $k \in \{2, 3, 4, 6\}$. They also determined the exact value of $\kappa^c_k(\overleftrightarrow{K}_{a,a})$ for each $2 \leq k \leq n$.

References

1. Aouchiche, M., Hansen, P.: A survey of Nordhaus-Gaddum type relations. Discrete Appl. Math. **161**(4/5), 466–546 (2013)
2. Bang-Jensen, J.: Locally semicomplete digraphs: a generalization of tournaments. J. Graph Theory **14**(3), 371–390 (1990)
3. Bang-Jensen, J.: Edge-disjoint in- and out-branchings in tournaments and related path problems. J. Combin. Theory Ser. B **51**(1), 1–23 (1991)
4. Bang-Jensen, J., Gutin, G.: Generalizations of tournaments: a survey. J. Graph Theory **28**, 171–202 (1998)
5. Bang-Jensen, J., Gutin, G.: Digraphs: Theory, Algorithms and Applications, 2nd edn. Springer, London (2009)
6. Bang-Jensen, J., Havet, F.: Tournaments and semicomplete digraphs. In: Bang-Jensen, J., Gutin, G. (eds.) Classes of Directed Graphs. Springer, Berlin (2018)
7. Bang-Jensen, J., Hell, P.: Fast algorithms for finding Hamiltonian paths and cycles in in-tournament digraphs. Discrete Appl. Math. **41**(1), 75–79 (1993)
8. Bang-Jensen, J., Huang, J.: Quasi-transitive digraphs. J. Graph Theory **20**(2), 141–161 (1995)
9. Bang-Jensen, J., Huang, J.: Decomposing locally semicomplete digraphs into strong spanning subdigraphs. J. Combin. Theory Ser. B **102**, 701–714 (2012)
10. Bang-Jensen, J., Huang, J.: Arc-disjoint in- and out-branchings rooted at the same vertex in locally semicomplete digraphs. J. Graph Theory **77**, 278–298 (2014)
11. Bang-Jensen, J., Wang, Y.: Arc-disjoint out- and in-branchings in compositions of digraphs. Eur. J. Combin. **120**, Article 103981 (2024)
12. Bang-Jensen, J., Yeo, A.: Decomposing k-arc-strong tournaments into strong spanning subdigraphs. Combinatorica **24**(3), 331–349 (2004)
13. Bang-Jensen, J., Frank, A., Jackson, B.: Preserving and increasing local edge-connectivity in mixed graphs. SIAM J. Discrete Math. **8**, 155–178 (1995)
14. Bang-Jensen, J., Guo, Y., Gutin, G., Volkmann, L.: A classification of locally semicomplete digraphs. Discrete Math. **167/168**, 101–114 (1997)
15. Bang-Jensen, J., Gutin, G., Yeo, A.: Arc-disjoint strong spanning subgraphs of semicomplete compositions. J. Graph Theory **95**(2), 267–289 (2020)
16. Bang-Jensen, J., Havet, F., Yeo, A.: Spanning Eulerian subdigraphs of semicomplete digraphs. J. Graph Theory **102**, 578–606 (2023)
17. Berczi, K., Kovacs, E.R.: A note on strongly edge-disjoint arborescences. EGRES TR-2011-04
18. Camion, P.: Chemins et circuits hamiltoniens des graphes complets. C. R. Acad. Sci. Paris **249**, 2151–2152 (1959)

Y. Sun, *Steiner Type Packing Problems in Digraphs*, SpringerBriefs in Mathematics,
https://doi.org/10.1007/978-981-95-8743-8

19. Chen, L., Li, X., Liu, M., Mao, Y.: A solution to a conjecture on the generalized connectivity of graphs. J. Combin. Optim. **33**(1), 275–282 (2017)
20. Cheriyan, J., Salavatipour, M.: Hardness and approximation results for packing Steiner trees. Algorithmica **45**, 21–43 (2006)
21. Chudnovsky, M. , Scott, A., Seymour, P.D.: Disjoint paths in unions of tournaments. J. Combin. Theory Ser. B **135**, 238–255 (2019)
22. DeVos, M., McDonald, J., Pivotto, I.: Packing Steiner trees. J. Combin. Theory Ser. B **119**, 178–213 (2016)
23. Dong, Y., Gutin, G., Sun, Y.: Strong subgraph 2-arc-connectivity and arc-strong connectivity of Cartesian product of digraphs. Discuss. Math. Graph Theory **44**(3), 913–931 (2024)
24. Edmonds, J.: Edge-disjoint branchings. In: Rustin, B. (ed.) Combinatorial Algorithms, pp. 91–96. Academic Press, New York (1973)
25. Feige, U., Halldorsson, M., Kortsarz, G., Srinivasan, A.: Approximating the domatic number. SIAM J. Comput. **32**(1), 172–195 (2002)
26. Fortune, S., Hopcroft, J., Wyllie, J.: The directed subgraphs homeomorphism problem. Theoret. Comput. Sci. **10**, 111–121 (1980)
27. Frank, A., Király, T., Kriesell, M.: On decomposing a hypergraph into k connected sub-hypergraphs. Discrete Appl. Math. **131**, 373–383 (2003)
28. Garey, M.R., Johnson, D.S.: Computers and Intractability: A Guide to the Theory of NP-completeness. Freeman, San Francisco (1979)
29. Georgiadis, L., Tarjan, R. E.: Dominator tree certification and divergent spanning trees. ACM Trans. Algorithms **12**(1), Article 11, 42 pp. (2015)
30. Grötschel, M., Martin, A., Weismantel, R.: Packing Steiner trees: further facets. Eur. J. Combin. **17**, 39–52 (1996)
31. Grötschel, M., Martin, A., Weismantel, R.: Packing Steiner trees: polyhedral investigations. Math. Program. **72**, 101–123 (1996)
32. Grötschel, M., Martin, A., Weismantel, R.: Packing Steiner trees: a cutting plane algorithm and computational results. Math. Program. **72**, 125–145 (1996)
33. Grötschel, M., Martin, A., Weismantel, R.: Packing Steiner trees: separation algorithms. SIAM J. Discrete Math. **9**, 233–257 (1996)
34. Grötschel, M., Martin, A., Weismantel, R.: The Steiner tree packing problem in VLSI design. Math. Program. **78**, 265–281 (1997)
35. Guo, Y., Volkmann, L.: Connectivity properties of locally semicomplete digraphs. J. Graph Theory **18**(3), 269–280 (1994)
36. Guruswami, V., Khanna, S., Rajaraman, R., Shepherd, B., Yannakakis, M.: Near-optimal hardness results and approximation algorithms for edge-disjoint paths and related problems. J. Comput. Syst. Sci. **67**(3), 473–496 (2003)
37. Gutin, G., Sun, Y.: Arc-disjoint in- and out-branchings rooted at the same vertex in compositions of digraphs. Discrete Math. **343**(5), 111816 (2020)
38. Hager, M.: Pendant tree-connectivity. J. Combin. Theory Ser. B **38**, 179–189 (1985)
39. Hammack, R.H.: Digraphs Products. In: Bang-Jensen, J., Gutin, G. (eds.) Classes of Directed Graphs. Springer, Berlin (2018)
40. Huck, A.: Disproof of a conjecture about independent branchings in k-connected directed graphs. J. Graph Theory **20**(2), 235–239 (1995)
41. Huck, A.: Independent branchings in acyclic digraphs. Discrete Math. **199**, 245–249 (1999)
42. Huck, A.: Independent trees and branchings in planar multigraphs. Graphs Combin. **15**, 211–220 (1999)
43. Ibaraki, T., Poljak, S.: Weak three-linking in Eulerian digraphs. SIAM J. Discrete Math. **4**(1), 84–98 (1991)
44. Jain, K., Mahdian, M., Salavatipour, M.R.: Packing Steiner trees. In: SODA, pp. 266–274 (2003)
45. Jeong, G.W., Lee, K., Park, S., Park, K.: A branch-and-price algorithm for the Steiner tree packing problem. Comput. Oper. Res. **29**, 221–241 (2002)

46. Johnson, T., Robertson, N., Seymour, P.D., Thomas, R.: Directed tree-width. J. Combin. Theory Ser. B **82**(1), 138–154 (2001)
47. Kriesell, M.: Edge-disjoint trees containing some given vertices in a graph. J. Combin. Theory Ser. B **88**, 53–65 (2003)
48. Kriesell, M.: Edge disjoint Steiner trees in graphs without large bridges. J. Graph Theory **62**, 188–198 (2009)
49. Kriesell, M.: Packing Steiner trees on four terminals. J. Combin. Theory Ser. B **100**, 546–553 (2010)
50. Kühn, D., Osthus, D.: Linkedness and ordered cycles in digraphs. Combin. Probab. Comput. **17**, 689–709 (2008)
51. Kühn, D., Osthus, D.: Hamilton decompositions of regular expanders: a proof of Kelly's conjecture for large tournaments. Adv. Math. **237**, 62–146 (2013)
52. Lau, L.: An approximate max-Steiner-tree-packing min-Steiner-cut theorem. Combinatorica **27**, 71–90 (2007)
53. Li, X., Mao, Y.: Generalized Connectivity of Graphs. Springer, Switzerland (2016)
54. Li, X., Mao, Y., Sun, Y.: On the generalized (edge-)connectivity of graphs. Australas. J. Combin. **58**(2), 304–319 (2014)
55. Lovász, L.: Coverings and colorings of hypergraphs. In: Proc. 4th Southeastern Conf. on Comb., Utilitas Math., pp. 3–12 (1973)
56. Nash-Williams, C.St.J.A.: Edge-disjoint spanning trees of finite graphs. J. Lond. Math. Soc. **36**, 445–450 (1961)
57. Ng, L.L.: Hamiltonian decomposition of complete regular multipartite digraphs. Discrete Math. **177**(1–3), 279–285 (1997)
58. Ng, L.L.: Hamiltonian decomposition of lexicographic products of digraphs. J. Combin. Theory Ser. B **73**(2), 119–129 (1998)
59. Pulleyblank, W.R.: Two Steiner tree packing problems. In: STOC, pp. 383–387 (1995)
60. Robertson, N., Seymour, P.: Graph minors XIII: the disjoint paths problem. J. Combin. Theory Ser. B **63**(1), 65–110 (1995)
61. Schrijver, A.: Finding k partially disjoint paths in a directed planar graph. SIAM J. Comput. **23**(4), 780–788 (1994)
62. Schrijver, A.: Combinatorial Optimization: Polyhedra and Efficiency. Springer, Berlin (2003)
63. Sherwani, N.: Algorithms for VLSI Physical Design Automation, 3rd edn. Kluwer Acad. Pub., London (1999)
64. Steiglitz, K., Weiner, P., Kleitman, D.: The design of minimum-cost survivable networks. IEEE Trans. Circuit Theory **16**(4), 455–460 (1969)
65. Sun, Y.: Compositions of digraphs: a survey. Discrete Appl. Math. **373**, 137–153 (2025)
66. Sun, Y.: Extremal results for directed tree connectivity. Bull. Malays. Math. Sci. Soc. **45**(2), 839–850 (2022)
67. Sun, Y., Gutin, G.: Strong subgraph connectivity of digraphs: a survey. J. Interconnection Netw. **21**(2), 2142004 (2021)
68. Sun, Y., Gutin, G.: Strong subgraph connectivity of digraphs. Graphs Combin. **37**(3), 951–970 (2021)
69. Sun, Y., Jin, Z.: Minimally strong subgraph (k, ℓ)-arc-connected digraphs. Discuss. Math. Graph Theory **42**(3), 759–770 (2022)
70. Sun, Y., Jin, Z.: Semicomplete compositions of digraphs. Discrete Math. **346**(8), Article 113420 (2023)
71. Sun, Y., Jin, Z.: Kings and kernels in semicomplete compositions, submitted (Also see: arXiv:2006.05607v4 [math.CO] for an earlier version)
72. Sun, Y., Jin, Z.: Perfect out-forest problem and directed Steiner cycle packing problem. Discrete Appl. Math. **366**, 201–209 (2025)
73. Sun, Y., Yeo, A.: Directed Steiner tree packing and directed tree connectivity. J. Graph Theory **102**(1), 86–106 (2023)

74. Sun, Y., Zhang, X.: Algorithmic and structural results of directed Steiner path packing and directed path connectivity. Graphs Combin. **42**(1), Article 2 (2026)
75. Sun, Y., Gutin, G., Ai, J.: Arc-disjoint strong spanning subdigraphs in compositions and products of digraphs. Discrete Math. **342**(8), 2297–2305 (2019)
76. Sun, Y., Gutin, G., Yeo, A., Zhang, X.: Strong subgraph k-connectivity. J. Graph Theory **92**(1), 5–18 (2019)
77. Sun, Y., Gutin, G., Zhang, X.: Packing strong subgraph in digraphs. Discrete Optim. **46**, Article 100745 (2022)
78. Thomassen, C.: Edge-disjoint Hamiltonian paths and cycles in tournaments. Proc. Lond. Math. Soc. **45**(1), 151–168 (1982)
79. Thomassen, C.: Configurations in graphs of large minimum degree, connectivity, or chromatic number. Ann. N. Y. Acad. Sci. **555**, 402–412 (1989)
80. Trotter, W.T., Erdős, P.: When the Cartesian product of directed cycles is Hamiltonian. J. Graph Theory **2**(2), 137–142 (1978)
81. Tutte, W.: On the problem of decomposing a graph into n connected factors. J. Lond. Math. Soc. **36**, 221–230 (1961)
82. Uchoa, E., Poggi de Aragão, M.: Vertex-disjoint packing of two Steiner trees: polyhedra and branch-and-cut. Math. Program. **90**, 537–557 (2001)
83. Wang, C., Sun, Y.: Directed cycle k-connectivity of complete digraphs and complete regular bipartite digraphs. Discrete Appl. Math. **358**, 203–213 (2024)
84. West, D., Wu, H.: Packing Steiner trees and S-connectors in graphs. J. Combin. Theory Ser. B **102**, 186–205 (2012)
85. Whitty, R.W.: Vertex-disjoint paths and edge-disjoint branchings in directed graphs. J. Graph Theory **11**, 349–358 (1987)

Index

Y. Sun, *Steiner Type Packing Problems in Digraphs*, SpringerBriefs in Mathematics,
https://doi.org/10.1007/978-981-95-8743-8

The manufacturer's authorised representative in the EU is Springer Nature Customer Service Centre GmbH, Europaplatz 3, 69115 Heidelberg, Germany. If you have any concerns regarding our products, please contact ProductSafety@springernature.com

Printed and bound by CPI Group (UK) Ltd, Croydon, CR0 4YY

07/07/2026

02160931-0003